Electronics Projects
for Beginners

Nikhil Shukla

V&S PUBLISHERS

Published by:

V&S PUBLISHERS

F-2/16, Ansari road, Daryaganj, New Delhi-110002
☎ 23240026, 23240027 • *Fax:* 011-23240028
Email: info@vspublishers.com • *Website:* www.vspublishers.com

Regional Office : Hyderabad
5-1-707/1, Brij Bhawan (Beside Central Bank of India Lane)
Bank Street, Koti, Hyderabad - 500 095
☎ 040-24737290
E-mail: vspublishershyd@gmail.com

Branch Office : Mumbai
Jaywant Industrial Estate, 1st Floor–108, Tardeo Road
Opposite Sobo Central Mall, Mumbai – 400 034
☎ 022-23510736
E-mail: vspublishersmum@gmail.com

Follow us on: t f in

Publisher's Note

V&S Publishers is pleased to bring out a new book – **Electronics Projects For Beginners**, designed for high school students. The practical work helps the students in understanding the concepts in a systematic and scientific way.

The book consists of various projects that help students to learn advanced scientific principles and develop skills in electronics field. Every project has been explained thoroughly describing all important aspects. The book includes topics such as FM Jammers, Mobile Jammers, Intruder Alarm, Room Thermostat, Car Parking Alarm, Sound Operated Light, Metal Detector, Educational Game, Audio Power Amplifier etc. The materials required to do projects are commonly available at home or are easily available at minimal cost in the market.

The basic idea behind publishing this book is to provide students with the study material which is interesting, educative, and practical. With the help of this book the students can easily do the given projects and can think of doing other projects also. The contents are written in a simple and lucid language for better understanding of the concepts and explanations.

Thus, to gain inventive skills you need to go through the given projects, which will definitely prove to be a great learning experience for all of you!

Contents

Electronic Projects

Introduction

Welcome to the world of experimentation and practical science around us. In fact, science is the knowledge that is obtained through reading, experimentation, observation and realization in a systematic manner. Any performance done systematically is said to have been done scientificall . In short, science means a system which is enjoyable, interesting, and thought-provoking.

Electricity and Electronics

There is a very little difference between electricity and electronics. Generally, Electronics is considered to be a branch of Electricity. It deals with electricity since electrons move through electricity and are affected by certain devices like resistors, capacitors, coils, transistors, and integrated circuits. The human race has harnessed electronic power to perform various tasks such as to illuminate a light bulb, make a calculating machine called computer which can do mathematics operations at tremendous speed. Electronics is among the most rapidly changing sciences as various technological advancements are made every year.

In homes today you will find a wealth of electronic marvels such as computers, stereos, televisions, radios, telephones, laser music discs, copy machines. Electrical appliances such as washing machines, hair dryers, vacuum cleaners, toasters, microwave ovens, electric can openers, refrigerators, air conditioners etc. have enhanced our standard of living.

This book includes many topics and projects based on electromagnetic forces, static electricity, flow of current, motors and generators, resistance and capacitances, generating electricity, solid state electronics, and radio frequency energy. You have to only make your selection, and begin your project. Most of the materials needed to do projects are commonly found at home or are easily available at minimal cost in the market.

Aim of Experimental Study

An experiment is the base of the development and growth of science. In science the aim of experimentation is to verify a given law which has already been derived from a theory. New discoveries can be made while doing an experiment with open eyes and attentive mind.

Significance of Practical Work in Laboratory

Physics is an experimental science. From the past events, we find that most of the path-breaking discoveries have been made by scientists while experimenting on an already known fact. Though the theory is taught in the class, performing an experiment by oneself makes the taught principles quite clear. It is just like learning by doing. Practical works done in a laboratory provide young minds the systematic and scientific training. To avoid laboratory woks is just like to learn to swim without going to the swimming pool.

Features of the Book

This book deals with a wide variety of topics related to Electricity and Electronics. The scientific concepts and projects introduced in the book will help the students to understand advanced scientific principles and develop many skills in science, which are required to sort out everyday problems in our ever-increasing complex society. All the projects are classified into three grades (A, B, and C) for beginners, intermediates, and professionals respectively. In each project, complete theory has been described systematically covering all important aspects. We suggest making a 'schematic diagram' of each project, which shows a pictorial layout of an electrical circuit and the arrangement of components and their interconnection. This will enhance your display visually and comprehension as well. The book includes various activities which are useful for beginners and professionals i.e. it covers from the sixth standard to the graduate level.

Project-

Analog Delay Timer

Can you build a project for devices i.e. compressors and halogen lamps which can not be OFF and ON repeatedly within a short period of time as it will cause damage to the devices ?

Introduction

In this delay timer project, all analog parts are being used with the thyristor as a device that switches an AC relay ON or OFF depending on the timing of the RC circuit. The input mains supply used ranges from 220V AC to 240 V AC and an AC relay (220-240V AC) is used to switch a load. The load to be switched must be within the current and relay ratings. This project is useful for the use of devices that need to be OFF for a minimum of 150-120 seconds after the mains supply has cut-off.

Materials Required

SCR	:	Thyristor 2P6M,MCR 106-8 or equivalent
$D_1 - D_9$	:	1N4001 diodes
$Q_1 - Q_2$	:	2SC2002 NPN transistor
Q_3	:	2SA953 PNP transistor
E_1	:	10 μF electrolytic capacitor
E_2	:	47 μF electrolytic capacitor
C_1	:	0.47 μF ceramic disc capacitor
RLY	:	AC relay
Z	:	9.1V zener diode
R_1	:	47 kΩ resistor
R_2	:	4.7 kΩ resistor
R_3	:	2.2 kΩ resistor
R_4	:	100 kΩ resistor
R_5	:	100 kΩ resistor
R_6	:	100 kΩ resistor
R_7	:	MΩ resistor
R_8	:	47 kΩ resistor
R_9	:	100 kΩ resistor
R_{10}	:	4.7 kΩ resistor

R$_{11}$:   4.7 kΩ resistor

1, 2, 3   :   Terminals

Assembly

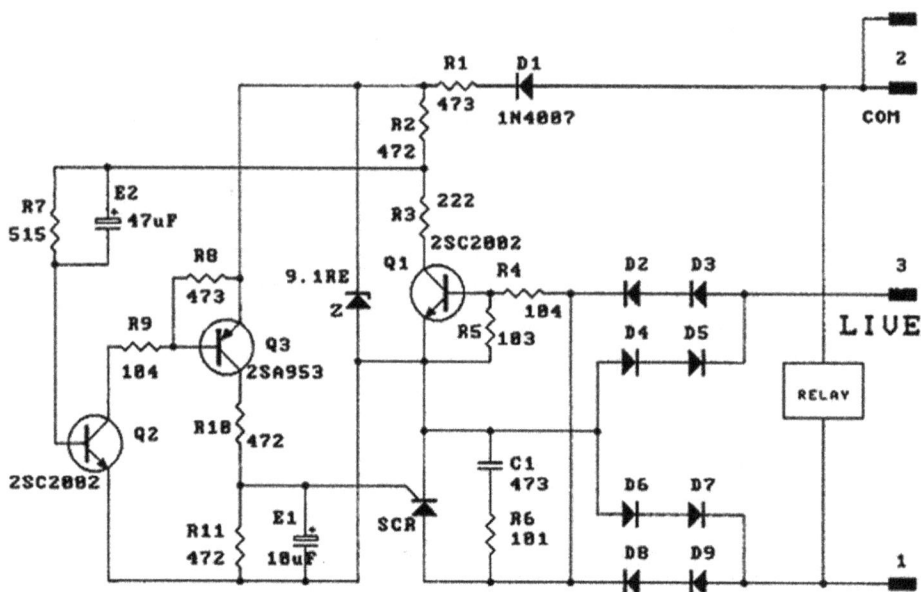

Fig. 1.1 : Circuit Diagram

The schematic fig. 1.1 shows the circuit diagram of the ON delay timer. Once the mains power supply cutt-off, the relay will only be able to turn ON after a period of 150-210 seconds depending on the tolerance of the RC circuit represented by resistor R$_7$ and electrotylic capacitor E$_2$. More accurate timing can be achieved by using low tolerance resistor and capacitor. The thyristor used can be either 2P6M or MCR106-8 or equivalent parts available in the market. Relays used should have coil rating below 1A in order not to overheat the SCR. No heat sink is required for SCR.

To Do and Notice

When the mains supply cuts-off, transistor Q$_2$ can not turn ON until whole charge of capacitor E$_2$ has been discharged through resistor R$_7$.

The duration for discharging the capacitor takes around 150-120 seconds. The timing of the circuit can be changed by reducing or increasing the RC values of R$_7$ and E$_2$.

What Happens?

At power ON, there is no charge on E$_2$, hence the transistor will be forward bais and turn ON when Q$_3$ turns ON.

Once these two transistors are ON, the SCR will turn ON as well. The use of C_1 and R_6 across the SCR acts as a snubber circuit to reduce the switching noise generated by the SCR when it turns OFF/ON. During the ON stage of the SCR, the capacitor E_2 is charged to its maximum value.

When mains supply cuts-off, the charge of E_2 will cause the base transistor Q_2 to be reverse bias and cannot turn ON until almost all the charges have been discharged through R_7. Once the charge has been discharged, transistor Q_2 will be able to turn ON.

Try It Yourself

Change the turn ON timing to 3 minutes and repeat the project.

Project-

Audio Power Amplifier

How to amplify an audio signal ?

Introduction

The audio power amplifier project is based on LM 1875 amplifier module. It is able to deliver up to 30W of power using an 8Ω load and dual 30V DC power supplies. It is design to operate with minimum external components with current limit and thermal shutdown protection features. Other features include high gain, fast slow rate, wide power supply range, large output voltage swing and high current capability.

Materials Required

U_1	:	LM 875 audio power amplifie
SPK_1	:	8 Ω speaker
T_1	:	240V/36V 80A centre topped transformer
D_1, D_2, D_3, D_4	:	1N5401 diode
V_1	:	275V diameter 14 mm varistor
S_1	:	Power switch
F_1	:	5A 240V AC fuse
F_2	:	2A 240V AC fuse
E_1, E_2, E_3, E_4	:	2200 µF/63V electrolytic capacitor
E_5	:	1 µF/63V electrolytic capacitor
E_6	:	22 µF/63V electrolytic capacitor
C_1, C_2, C_3, C_4	:	0.1 µF/50V ceramic capacitor
C_5	:	0.22 µF Polyester capacitor
R_1	:	1 µF 1/4W carbon film resisto
R_2	:	1 MΩ 1/4W carbon film resisto
R_3	:	22 kΩ 1/4W carbon film resisto
R_4	:	20 kΩ 1/4W carbon film resisto
R_5	:	1 kΩ 1/4W carbon film resisto
R_6	:	1 kΩ 1/4W carbon film resisto
Heat Sink	:	Termal resistrance of 1/4" C/W or above screw, nut, washer to mount heat sink to LM1875 insulated TO-220 washer

Assembly

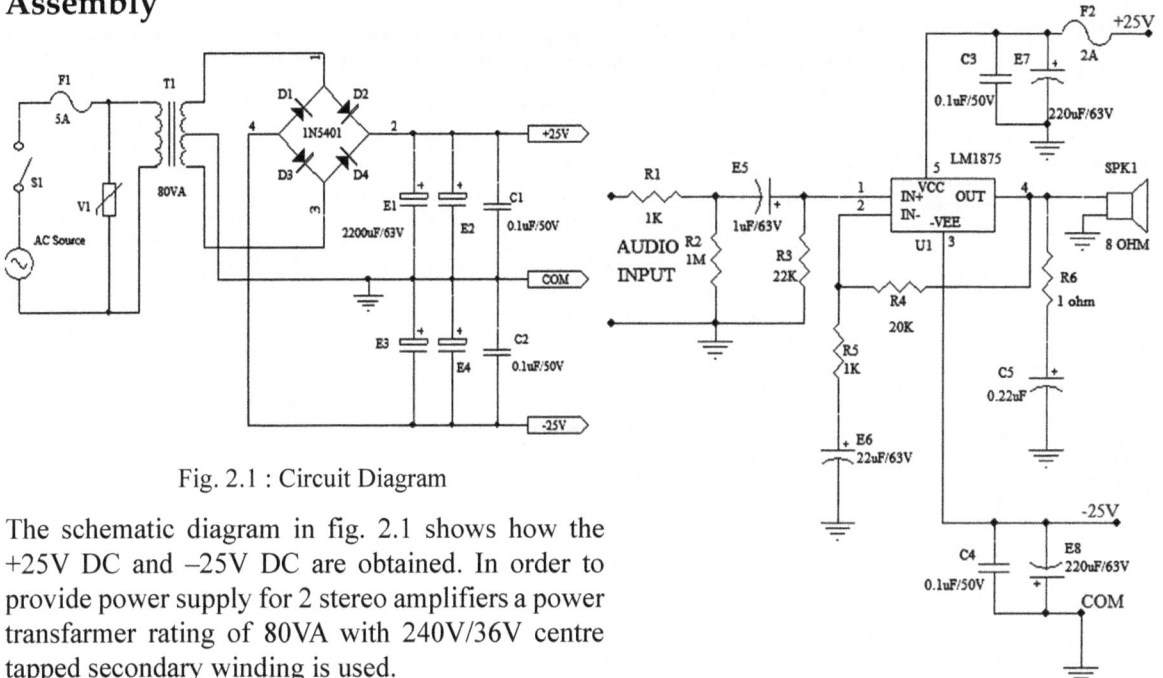

Fig. 2.1 : Circuit Diagram

The schematic diagram in fig. 2.1 shows how the +25V DC and –25V DC are obtained. In order to provide power supply for 2 stereo amplifiers a power transfarmer rating of 80VA with 240V/36V centre tapped secondary winding is used.

The secondary output of the transformer is rectified by using four 1N 5401 diodes together with 4 electrolytic capacitor to smoothen the riple voltage. A fuse and varistor are connected at the primary input to protect the circuit against power surge.

Fig. 2.2 : Circuit Diagram

To Do and Notice

When you give an audio signal at input terminal, we get output at speaker i.e. amplified audio signal.

What Happens?

The +25V and –25V DC power supply are connected to the audio amplifier module through a 2A fuse with the peripheral devices shown in the the fig 2.2.

The audio input signal to be amplified is coupled to pin 1 of LM 1875 through the resistor R_1 and electrolytic capacitor E_5. The output signal at pin 4 of LM 1875 can be used to directly drive a 8Ω loudspeaker. Resister R_6 and capacitor C_5 prevent the capacitance developed at the long speaker leads from driving the amplifier into very high frequency oscillation

A heat sink with a thermal resistance rating of 1.4°C/W or better must be used or else the amplifier module will be cut-off from operation due to the heat that will build up during the operation of the amplifie . Take note that the heat sink tab on the IC module is internally connected to the −25V power supply; hence, it must be isolated from the heat sink by the use of an insulating washer. If this is not done, the negative trail will be started to ground.

Try It Yourself

Make a categorigation table of all types of amplfiers

Project-

Auto Shut-off Tone Generator

Can you build an emergency siren which is shut off until the duration of the set has elapsed?

Introduction

In this auto shut-off tone generator project, once the switch to the 9V power supply is connected, the alarm will trigger at a frequency of approximately 1.27 kHz.

It will remain ON for a duration of approximately 2.8 minutes before it stopped. This project is useful when built as one can carry it along wherever one goes or places it in a vehicle. In times of emergency, one can easily switch ON the switch and the loud speaker will emit a loud sound that will frighten the uninvited guest.

Materials Required

U_1, U_2	:	NE 555 timer
R_1	:	18 kΩ 1/4W, 5% carbon film resisto
R_2	:	330 kΩ 1/4W, 5% carbon film resisto
R_3	:	47 kΩ 1/4W, 5% carbon film resisto
R_4	:	33 kΩ 1/4W, 5% carbon film resisto
R_5	:	220 kΩ 1/4W, 5% carbon film resisto
E_1	:	4.7 µF/25 V electrolytic capacitor
E_2	:	470 µF/25 V electrolytic capacitor
C_1	:	10 nF/25V ceramic capacitor
Q_1	:	TIP41A
SPKR	:	8Ω speaker
S_1	:	switch
BAT	:	9 V battery, battery holder

Assembly

The schematic of the tone generator is as shown the fig 67.1. It is based on two 555 timer ICs or one single 556 timer IC (which contains two 555 timers). In this schematic, two 555 timers are used. V2 is configured as a timer in astable mode

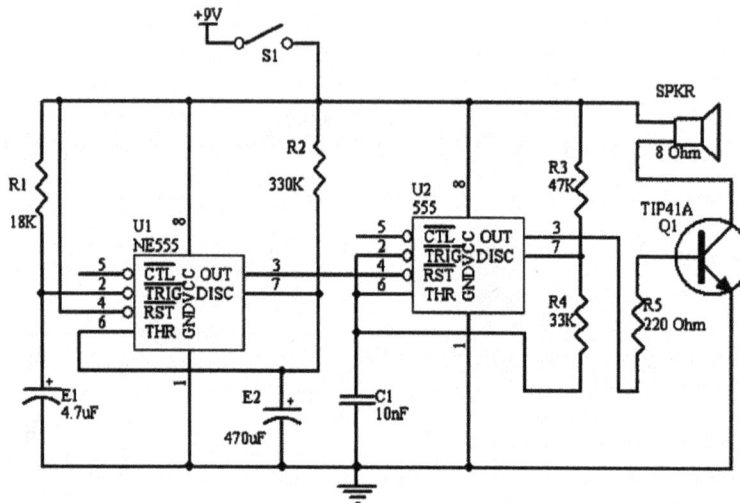

Fig 3.1 : Circuit Diagram

To Do and Notice

The pulse duration of the monostable circuit is given by the formula :

$$T = 1.1 \,(330 \text{ k}\Omega)\,(470 \text{ μF})$$
$$= 170 \text{ seconds}$$

Once this timing is up, the pulse output will disable the astable circuit U_2. Hence the loud sound emitted by the 8 Ω loud speaker stops emitting sound.

What Happens?

Once triggered it will emit a frequency from its outputs at Pin 3 that will drive a Q_1 transistor. Q_1 transistor will turn ON and OFF according to the frequency circuit. It will in turn be used to drive 8Ω loud speaker to emit a loud audible sound.

The astable frequency of circuit U_2 is given by the formula of 555 timer as shown below :

$$f = 1.44/[47 \text{ k}\Omega + 2\,(33 \text{ k}\Omega)]\,[10 \text{ nF}]$$
$$= 1.27 \text{ kHz}$$

The frequency of the sound can be adjusted by changing the values of $R_3 = 47\Omega$, R4 = 33Ω and capactor $C_1 = 10$ nF. Change the values of these components and by using the formula for astable mode, the frequency of the sound can be obtained.

U_1 circuit is used as a delay circuit which is configured as a monostable mode. It is one shot multivibrator that will generate a pulse at its output Pin 3 which will disable the astable circuit U_1. In this circuit, Pin-2 of U_1 will go to logic '0' when the power supply is connected via the capacitor C_1 and hence circuit U_2 is immediately triggered.

Try It Yourself

Increase the astable frequency of circuit U_2 in the above project to 2.54 kHz ?

Project-

Auto Shut-off Tone Generator

Can you build an emergency siren which is shut off until the duration of the set has elapsed?

Introduction

In this auto shut-off tone generator project, once the switch to the 9V power supply is connected, the alarm will trigger at a frequency of approximately 1.27 kHz.

It will remain ON for a duration of approximately 2.8 minutes before it stopped. This project is useful when built as one can carry it along wherever one goes or places it in a vehicle. In times of emergency, one can easily switch ON the switch and the loud speaker will emit a loud sound that will frighten the uninvited guest.

Materials Required

U_1, U_2	:	NE 555 timer
R_1	:	18 kΩ 1/4W, 5% carbon film resisto
R_2	:	330 kΩ 1/4W, 5% carbon film resisto
R_3	:	47 kΩ 1/4W, 5% carbon film resisto
R_4	:	33 kΩ 1/4W, 5% carbon film resisto
R_5	:	220 kΩ 1/4W, 5% carbon film resisto
E_1	:	4.7 µF/25 V electrolytic capacitor
E_2	:	470 µF/25 V electrolytic capacitor
C_1	:	10 nF/25V ceramic capacitor
Q_1	:	TIP41A
SPKR	:	8Ω speaker
S_1	:	switch
BAT	:	9 V battery, battery holder

Assembly

The schematic of the tone generator is as shown the fig 4.1. It is based on two 555 timer ICs or one single 556 timer IC (which contains two 555 timers). In this schematic, two 555 timers are used. V2 is configured as a timer in astable mode

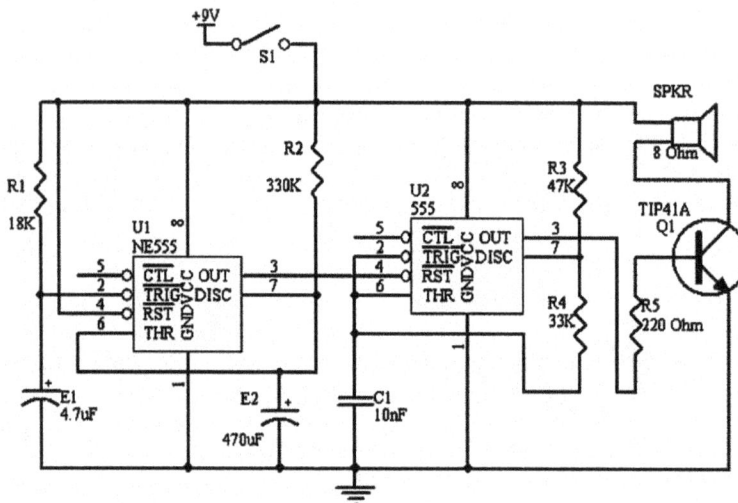

Fig 4.1 : Circuit Diagram

To Do and Notice

The pulse duration of the monostable circuit is given by the formula :

$$T = 1.1 \ (330 \ k\Omega) \ (470 \ \mu F)$$
$$= 170 \ seconds$$

Once this timing is up, the pulse output will disable the astable circuit U_2. Hence the loud sound emitted by the 8 Ω loud speaker stops emitting sound.

What Happens?

Once triggered it will emit a frequency from its outputs at Pin 3 that will drive a Q_1 transistor. Q_1 transistor will turn ON and OFF according to the frequency circuit. It will in turn be used to drive 8Ω loud speaker to emit a loud audible sound.

The astable frequency of circuit U_2 is given by the formula of 555 timer as shown below :

$$f = 1.44/[47 \ k\Omega + 2 \ (33 \ k\Omega)] \ [10 \ nF]$$
$$= 1.27 \ kHz$$

The frequency of the sound can be adjusted by changing the values of $R_3 = 47\Omega$, R4 = 33Ω and capactor $C_1 = 10$ nF. Change the values of these components and by using the formula for astable mode, the frequency of the sound can be obtained.

U_1 circuit is used as a delay circuit which is configured as a monostable mode. It is one shot multivibrator that will generate a pulse at its output Pin 3 which will disable the astable circuit U_1. In this circuit, Pin-2 of U_1 will go to logic '0' when the power supply is connected via the capacitor C_1 and hence circuit U_2 is immediately triggered.

Try It Yourself

Increase the astable frequency of circuit U_2 in the above project to 2.54 kHz ?

Project-

Boolean Algebra Calculator

How to design a circuit which calculates the Boolean algebra problems ?

Introduction

The Boolean algebra calculator gives a basic logic for the operations on binary numbers 0 and 1. This boolean calculator is well known as boolean algebra simplifier or boolean algebra solver. In our circuit, we use Boolean algebra simplification methods like the Quine McCluskey algorithm to simplify the Boolean expression and display the output.

It works as a portable calculator to simplify the Boolean expression on the fl .

Materials Required

$R_1, R_2, R_3, R_4, R_5, R_6, R_7, R_8, R_9$	:	3.3 kΩ carbon film resisto
R_{11}	:	1 kΩ
R_{10}	:	10 kΩ variable resistor
$S_1, S_2, S_3, S_4, S_5, S_6, S_7, S_8$	:	ON/OFF switch
$LED_1, LED_2, LED_3, LED_4, LED_5, LED_6, LED_7$:		LEDs
IC_1	:	7805 voltage regularor
Battery	:	9V battery (DC)
IC_2	:	ATMEGA 16 L
		microcontroller

Assembly

Fig. 5.2 shows the block diagram of Boolean expression minimization calculator. Block diagram shows the 4 parts of our project which are

(i) **Power supply :** In our project a supply mains that is 5V DC is given to micro controller, LEDs, keypad and display.

(ii) **Micro controller :** ATMEGA 16 L is used for the automation purpose and acts as brain of the project. It controls the output (display) according to the input given to it.

(iii) **Display :** The display used here is 3 Bi-colour LEDs. The glowing pattern of LEDs represent the desired minimized expresion.

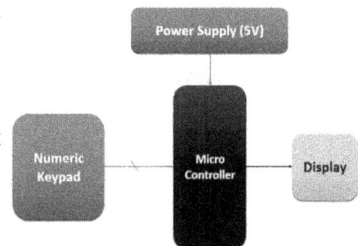

Fig. 5.1 : Power Supply

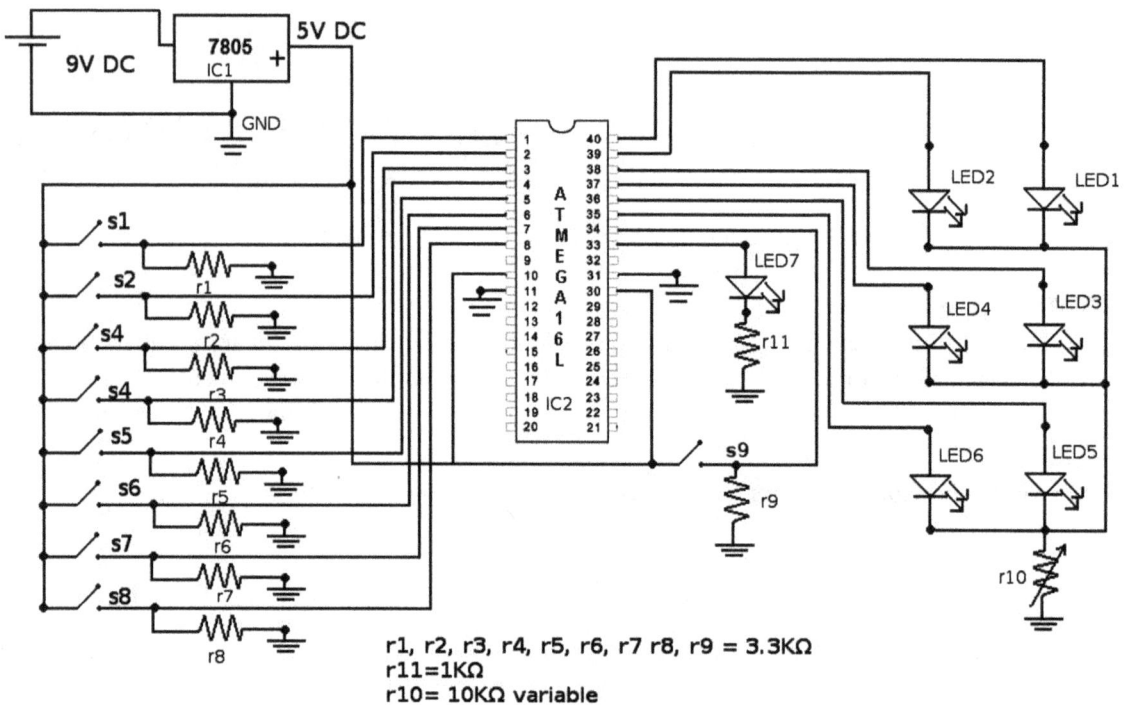

r1, r2, r3, r4, r5, r6, r7 r8, r9 = 3.3KΩ
r11=1KΩ
r10= 10KΩ variable

Fig. 5.2 : Circuit Diagram

(iv) **Keypad :** In this project a series of switches have been used as a keypad. It is used to give the input (minterms) expression.

Each digit on the keypad corresponds to one mintern. Fig. 5.2 shows the main circuit diagram of the project.

To Do and Notice

The output is displayed as one minterm at a time. Next minterm is displayed by pressing the next button and after reaching the last minterm of the reduced expression the input indicating LEDs is switched OFF which represents AND of the output. After few seconds it is again switched ON automatically when microcontroller is ready to take next input.

What Happens?

This circuit is a simple 3 variable Boolean expression minimizer. It uses the Quine McCluskey algorithm which was described. In this circuit micro-controller consist of code to implement the described algorithm and control other components in the circuit. Initially, when the power is switched ON, an LED will glow which indicates that the microcontroller is ready to take the input. Here, the input boolean expression is given in SOP form, i.e. only minterms are to be entered through the keypad provided.

The keypad consist of 9 switches of which 8 switches corresponds to one minterm each and the 9th switch is used as next button. After entering the expression the input indicating LED will go OFF. Now, based on the algorithm, microcontroller reduces the expresion the input representing LED glows which means that the expression has minimized and is displayed. The display consists of 3 Bi-colour LEDs in which green light represents the variable in normal form and the rest red light represents the variable in the complemented form, the order of them is as shown in the circuit diagram.

Try It Yourself
Understand PIC microcontroller and its design.

Project-

Communicating with Laser Beam

How can a given electrical signal be transmitted from the one place to another riding over a laser beam ?

Introduction

We are all familiar with fiber optic cables that carry our telephone signals from one place to another. An optical fiber is a flexible, transparent fibre made of high quality extruded glass (silica) or plastic, slightly thicker than a human hair. It transmits light between the two ends of the fibre. The laser bream here is used as a carrier, which is modulated by the signal to be transmitted.

The basic principle of operation is the same whether we have the fibre optic link or the wireless link. The wireless laser communication links are very popular in space applications for providing inter-satellite communications.

Materials Required

R_1	:	4.7 KΩ, 1/4W carbon film resisto
R_2	:	2.2 KΩ, 1/4W carbon film resisto
R_3	:	1.2 KΩ, 1/4W carbon film resisto
R_4	:	33 KΩ, 2W carbon film resisto
R_5	:	470 KΩ, 1/4W carbon film resisto
R_6	:	820 KΩ, 1/4W carbon film resisto
R_7	:	68 KΩ, 1/4W carbon film resisto
R_8, R_9	:	1.5 KΩ, 1/4W carbon film resisto
$P_1, P_2,$ and P_3	:	10 KΩ presets potentiometers
$C_1, C_2, C_3, C_4, C_5, C_6, C_7, C_8, C_9$	:	0.1 μF (ceramic disc) capacitors
C_2	:	0.1 μF (polyster/mica) capacitors
D_1, D_2	:	1N4001 diodes
LD-1	:	Laser pointer laser diode/any laser diode with about 1mW output power
PD-1	:	BPX-65 photodiode
IC-1	:	555 timer IC
IC-2, IC-3, IC-4	:	LF 356
B-1 to B-4	:	9V batteries
SW-1, SW-2	:	DPDT switch

Assembly

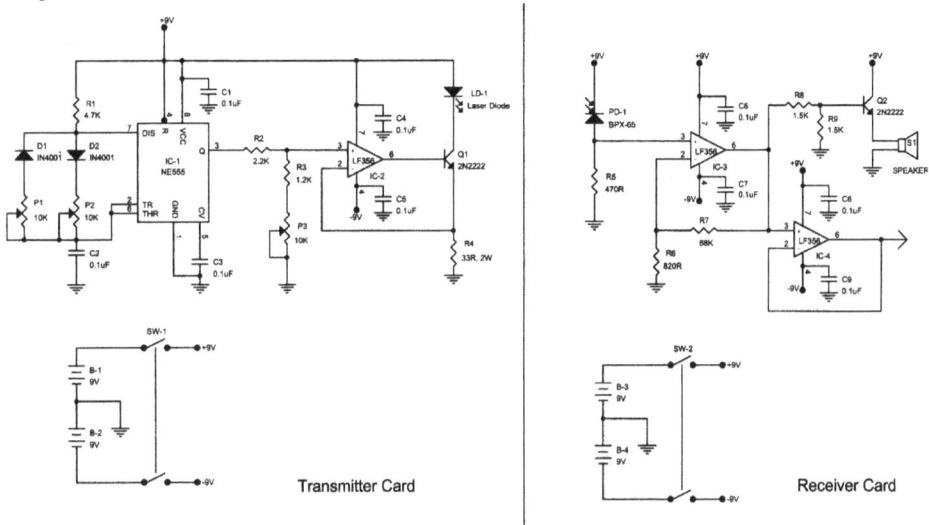

Fig. 6.1 : Circuit Diagram

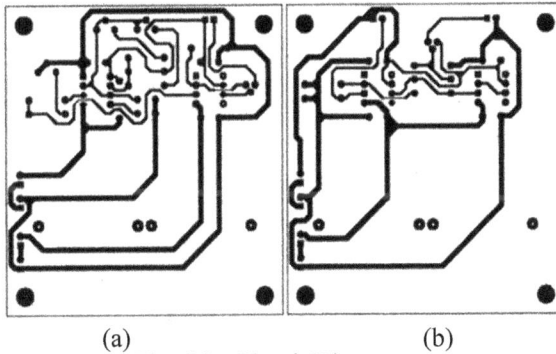

(a)　　　　　　　　　　　(b)

Fig. 6.2 : Circuit Diagram

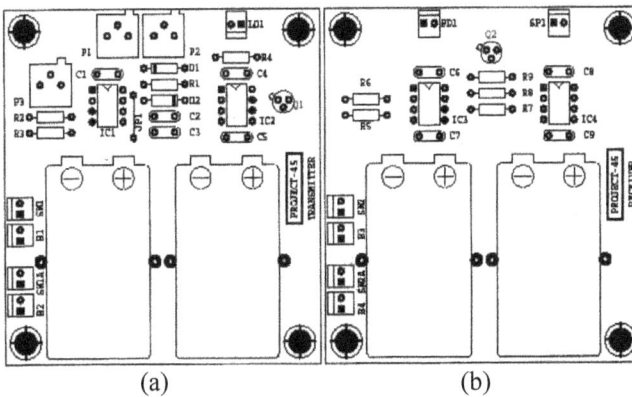

(a)　　　　　　　　　　　(b)

Fig. 6.3 : Circuit Diagram

Fig 6.1 shows the circuit diagram. The circuit is broadly divided into transmitter part and receiver part.

Fig. 6.2 (a) and (b) respectively shows the PCB and components layout of the transmitter circuit. The same for receiver circuit are shown in fig 6.3 (a) and (b). The two circuits are however simple enough to be assembled on general purpose PCBs.

To Do and Notice

The laser beam used is modulated by the signal to transmitted and it receives at receiver end the desired signal is separated from the carrier by demodulation technique.

What Happens?

The transmitter circuit comprises an astable multivibrator generating pulse train which serves as modulation input for the laser diode circuit. The output waveform appearing at Pin-3 of this IC has a high-time given by $0.69* (R_1 + P_2) *C_2$ and low time given by $0.69* P_1 *C_2$. The frequency of this pulse train can be set to the around $1kH_z$. The potential divider arrangement of R_2, R_3 and P_3 reduces the peak amplitude of the pulse train from about 8 volts to 3 volts. The laser diode current switches from zero to 90 mA.

The receiver part is current to voltage convertor followed by a non-inverting gain stage. The gain stage is built around IC-3 with gain value given by $[(R_6 + R_7)/R_6]$. The gain for this stage can be chosen so as to get sufficient signal aptitude at its output. The output drives a mini speaker (S_1) through an emitter follower arrangement configured around Q_2. The unity gain buffer stage built around IC-4 is to facilitate viewing of received signal on oscilloscope if desired so.

The receiver and transmitter circuits both are powered by two 9V batteries to provide +9V and –9V outputs. These voltages can be connected through DPDT switch SW-2 for receiver and SW-1 for transmitter circuits.

Try It Yourself

Make laser-based perimeter protection system.

Project-

Cooler Pump Protection cum Humidity Control

Can you protect your cooler pump and control humidity ?

Introduction

The desert cooler pumps pack off mainly due to exessive heating during summer peaks. Another short coming of these coolers is excessive humidity. Yet another problem faced particularly with low water capacity coolers is that you may suddenly discover at past mid night that there is no water in the cooler tank. All these problems could be overcome if the cooler pump is operated in a duty cycle while the fan was run continuously. Typically, the pump may be kept ON for two minues and then switched OFF for a time ranging from four to eight minutes depending upon environmental temperature. This not only enhances pump life but also controls humidity. In fact, with humidity control, cooling effect is better.

Materials Required

R_1, R_2	:	1 MΩ, 1/4 W carbon film resisto
R_3	:	2.2 kΩ, 1/4 W carbon film resisto
P_1	:	2.2 kΩ, potentiometer (LIN)
C_1	:	220 μF, 25V electrolytic capacitor
C_2	:	0.01 μF, 25V ceramic disc capacitor
C_3	:	1000 μF, 25V electrolytic capacitor
C_4	:	0.1 μF ceramic disc capacitor
D_1, D_2, D_3	:	IN4001 or equivalent diodes
VR-1	:	Three terminal regulator, type 7812
IC-1	:	555 timer
Relay	:	12V DC relay with at least one normally open contact
T-1	:	12V, 250 mA mains transformer
Fuse	:	0.25 A rating
S-1	:	Simple ON/OFF toggle switch

Assembly

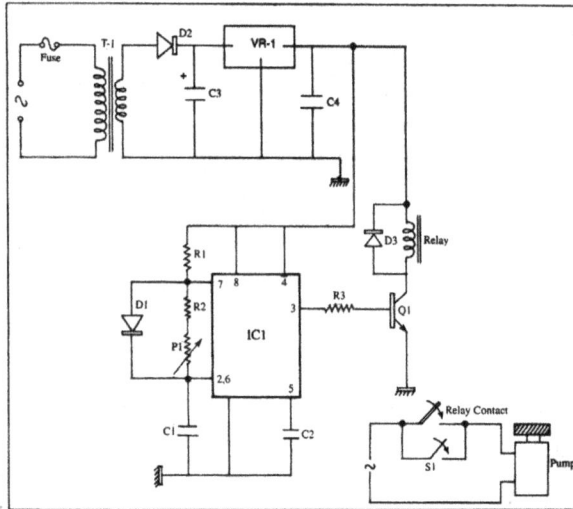

Fig. 7.1 : Circuit Diagram

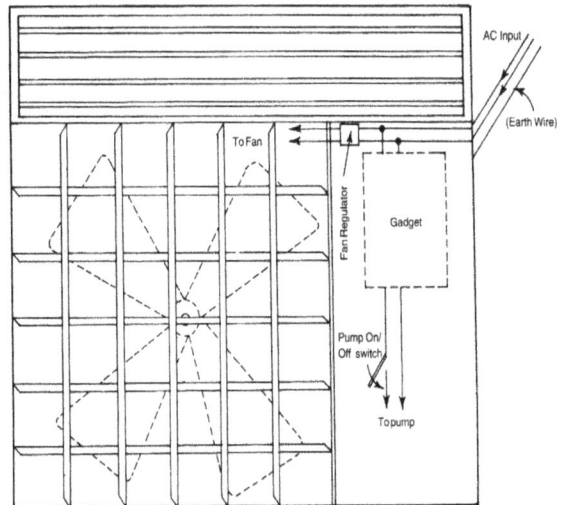

Fig. 7.2 : Circuit Diagram

As shown in fig. 7.1 the circuit is simple. It is built around IC timer 555 wired as an astable multivibrator whose high timer drives the relay ON and whose low time switches it OFF. The cooler fan gets the AC input directly whereas the cooler pump gets it via relay's normally open contact.

Fig. 7.2 shows the interconnections of the gadget and the cooler. The AC input for the cooler (also going to the cooler fan through the regulator) also goes to the input of the gadget. The assembled unit should be mounted on the inside of the front panel near the usual controls of the cooler. The cabinet should be the adequately sealed with some epoxy such as Araldite so that moisture does not get inside the unit.

To Do and Notice

We see that for 2 minutes the pump remains OFF which will variable from 3 minutes to 8 minutes. Switch S-1 acrosss the relay contact can be used in case you desire to run water pump continuously.

What Happens?

When the 555 timers output is high, the transistor conducts and the relay is energised. Normally open contact closes and the pump gets AC input. When the output goes low, transistor goes to cut-off, the relay is de-emergised and the pump is OFF. The high time of 555 output waveform is given by $(0.69 \, R_1 \, C_1)$ and the low time is given by $[0.69 \, (R_2 + P_1)C_1]$.

The given component values produce a fixed high time of approximately 3 to 8 minutes. Switc S–1 is used to bypass the relay contact in case you want to run the pump continuously.

Try It Yourself

Make a gadget which cuts-off the water pump when the water tank goes empty.

Project-

Digital Dice

Can you display in random the number from 1 to 9 on the 7 segment display?

Introduction

The digital dice project is an interesting project that will display in random the number 1 to 9 on the 7 segment display. This is an alternative device that can be used to replace the traditional dice when you are playing games such as Snake Ladder and Monopoly. This circuit is mostly for exercising and fun.

Materials Required

U_1	:	7490 decade counter
U_2	:	7483 4-bit binary adder
U_3	:	7447 BCD to 7 segment decoder
U_4	:	Common anode 7 segment display
U_5	:	555 timer
R_1, R_2	:	1kΩ, 1/4W, 1% resistor
R_3, R_9	:	270Ω, 1/4W, 5% resistor
PB	:	Normally open push button
C_1, C_2	:	0.01 µF/25V ceramic capacitor
E_1	:	220 µF/16V electroyltic capacitor

Assembly

The generation of clock is done by using a 555 timer which is connected in the astable mode at a frequency of approximately 50 Hz. The clock signal is fed into the decade counter which outputs are connected to 4 bit binary adder which provides a binary output equivalent to binary input +1. The outputs are then connected to a BCD to 7 segment decoder which is used to drive a common anode 7 segment display.

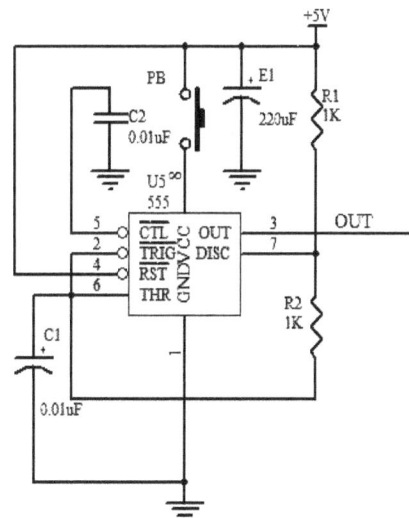

Fig 8.1 : Circuit Diagram

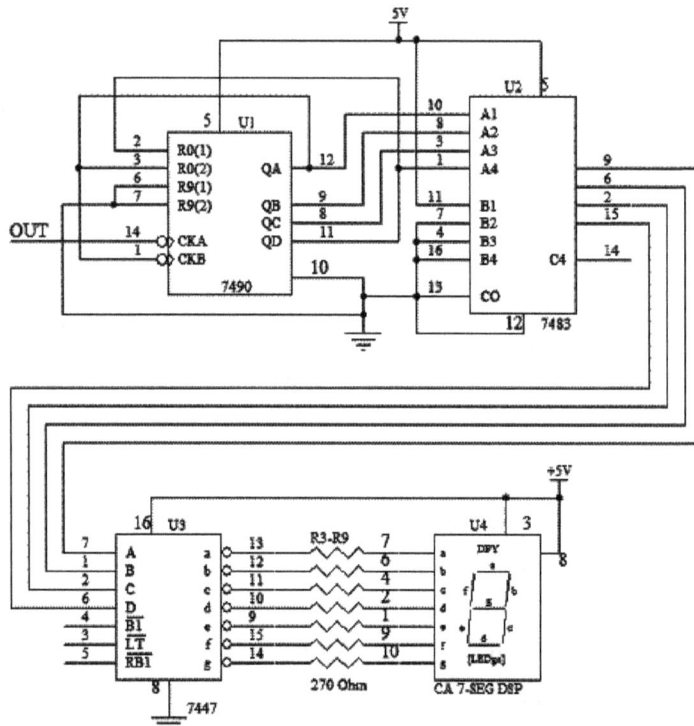

Fig 8.2 : Circuit Diagram

To Do and Notice

When push button PB is pressed, a square output will be generated from 555 timer which gives a frequency of approximately 50 Hz. to the 7490 decade counter IC. The frequency of the astable 555 timer is calculated by using the standard formula of the timer

$$f \quad = \quad 1.44/[(1k + 2 * 1k)(0.01 \ \mu F)]$$
$$= \quad 48 \ Hz.$$

What Happens?

Output from 555 timer is connected to the input of U_1 7490 decade counter when the decade counter reach the count of 9, the outputs of QA and QD will go to logic '1' and the counter is reset. The 7447 BCD to 7 segment decoder is used to drive the 7 segment common anode display.

Try It Yourself

Know the working of 555 timer in its monostable mode.

Project-

Educational Game

Can you make a project which is used in a quiz competition to see which of the participants press the button first to answer questions posed by the quiz master

Introduction

Quiz games are becoming very popular these days. The main concept behind these games is the use of fastest finger first indicator which are used to test the reaction time of the contestants. The project here has a circuit that can be used for 2 players. It can be used to identify which of the two has first pressed the button and locks out the other. This constructional project is based on a latch CD74HC75 IC and a 4 input CD74AC 20 NAND IC. The first person to press the switch will light up the LED designated for the person. The rest of theLEDs will not be activated until the RESET button is pressed.

Materials Required

U_1	:	CD74HC 75 or equivalent
U_2	:	CD74AC 20 or equivalent
R_1, R_2, R_3, R_4, R_5	:	1.5 kΩ 1/4W, 5% carbon film resisto
R_6, R_7, R_8, R_9	:	470 kΩ 1/4W, 5% carbon film resisto
$LED_1, LED_2, LED_3, LED_4$	:	5 mm or 3 mm LED
RESET, S_1, S_2, S_3, S_4	:	Normally open push button
C_1	:	0.1 μF/25V ceramic capacitor

Assembly

The truth table of the latch IC and the NAND IC is as shown in the figures belo .

Inputs		Outputs	
D	E	Q	$\overline{Q}$
L	H	L	H
H	H	H	L
X	L	Q_o	$\overline{Q}_0$

Fig. 9.1 : CD74HC 75 Latch Function Table

Inputs				Outputs
A	B	C	D	Y
H	H	H	H	L
L	X	X	X	H
X	L	X	X	H
X	X	L	X	H
X	X	X	L	H

Fig. 9.2 : NAND Gate Function Table

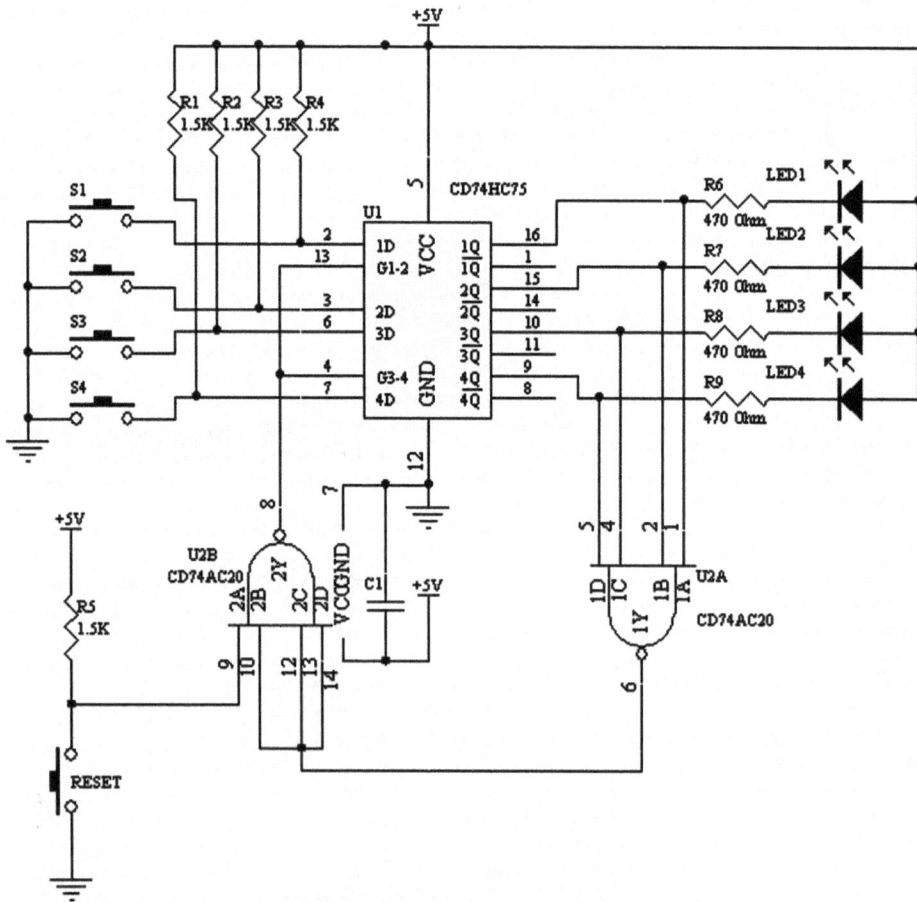

Fig. 9.3 : Circuit Diagram

To Do and Notice

If the first button pushed S_1 is a result of the player who pressed S_1, it will be shown as the one who has first pressed the button and so qualify to answer the question

Now the RESET button will be pressed by the quiz master to prepare the circuit for the next game.

What Happens?

Initially when RESET button is pressed, the output of NAND gate from U_2A will be logic '0' as all the inputs are at logic '1' state. As shown in the truth table for NAND gate, when all the inputs are '1', its output will be '0'. Hence, the output of NAND gate from U_2B will be logic '1' since an input of '0' will cause it to go to logic '1'. The logic '1' to the latch indicates that it is ready to output any data once any of the button is pushed.

If the first button pushed is S_1, the logic at its output Q_1 will be logic '0' and cause LED_1 to light up. At the same time, it causes the output of U_2A to go into logic '1'. Hence, the output of U_2B will be logic '0' causing the latch to be disable which means that the outputs of all the latches will be at their previous state. As a result, the player who pressed S_1 will be shown as the one who is first to press the button.

Try It Yourself

Build a project which calculates the total scores of two different teams.

Project-

Electromechanical Counter

Can you build a project which can be used to count the number of bottles passing over the conveyer belt or even the number of people passing through a certain entrance?

Introduction

In this project, the transmitter and receiver portions are aligned with each other and infra-red light beam generated by the transmitter is directly falling on the photo diode in the receiver. Whenever the infra-red light beam is interrupted, the electromechanical counter beam advances by one count. In a typical application of counting of bottles passing over a moving conveyer belt, the transmitter and receiver portions are located on the two opposite sides of the belt and aligned with each other to establish a continuous light bream path. The light path is interrupted every time a bottle moves across it. The number of bottles passing over a period of time is thus known from counter reading.

Materials Required

R_1	:	68Ω, 1W resistor
R_2	:	10MΩ, 1/4W resistor
R_3	:	1KΩ, 1/4W resistor
R_4	:	470Ω, 1/4W resistor
R_5	:	4.7Ω, 1/4W resistor
D_1	:	1N4001 or equivalent diode
D_2, D_3	:	Infra-red emitting diode type TIL34
D_4	:	Photo diode, type TIL81
D_5	:	LED
Q_1, Q_2	:	2N2222
IC-1	:	LM324 (quad opamp)
S_1, S_2	:	two ON/OFF switches

Two 9 V batteries, and electromechanical incremental counters (14 pin DIL IC socket)

Assembly

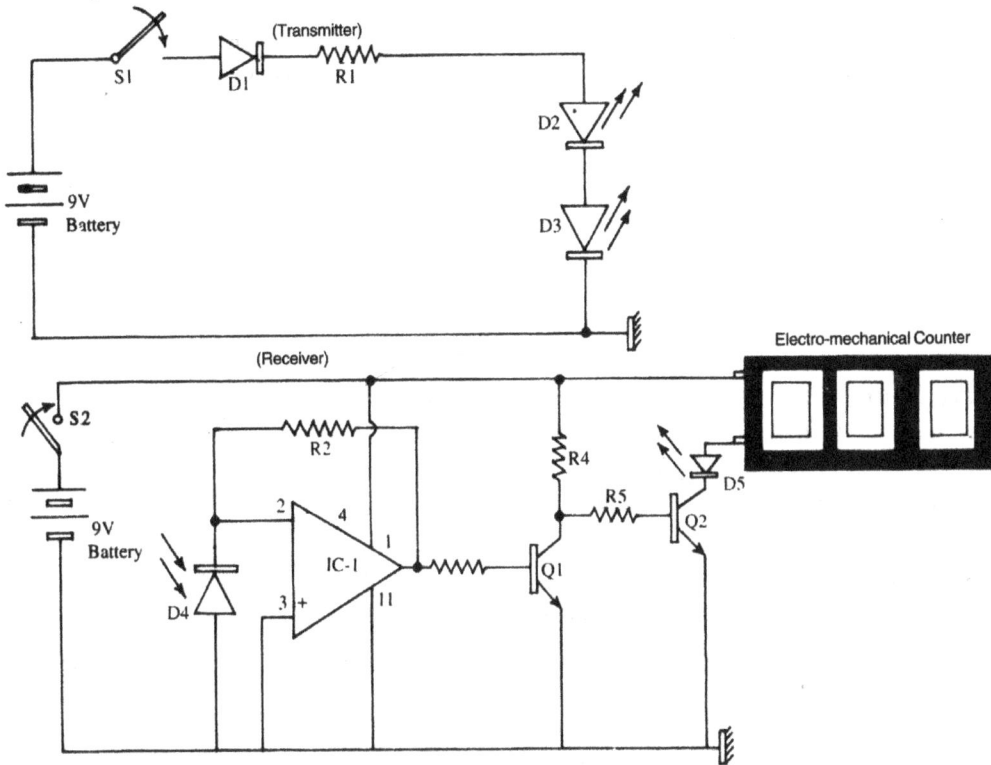

Fig. 10.1 : Circuit Diagram

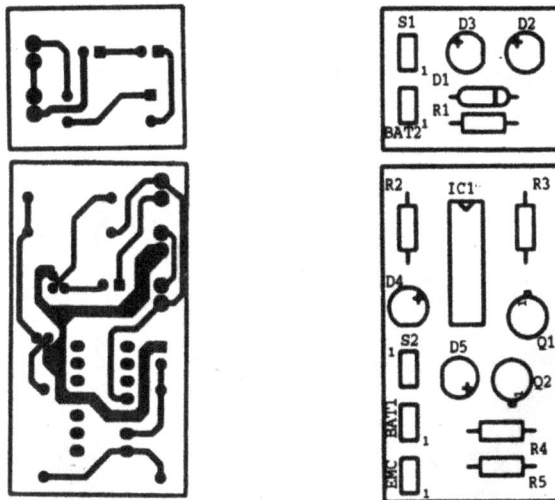

Fig. 10.2 PCB Layout and Components Layout

This gadget comprises a transmitter module and a receiver module. The transmitter module generates the required current drive for the infra-red light emitting diode and the receiver module once aligned with the transmittter module senses any interruption of the light beam caused due to any moving obstacle and generates an electrical pulse which advances the electromechanical counter by one count.

Fig. 10.1 and 10.2 respectively show the PCB layout and components layout.

To Do and Notice

When the infra-red light bream is blocked. The electromechanical counter increments by one count. If the light beam remains blocked continuously which does not happen in the intended application of the circuit, the counter would keep on incrementing at a steady rate.

What Happens?

In transmitter module, the current through the LED is being controlled by R_1. The current has been set at 60 mA. Two identical infa-red diodes have been connected in series to increase the light output to facilitate a larger range between transmitter and receiver locations.

In the receiver module, which during normal operation is aligned to the transmitter, the infa-red beam emitted by diodes D_2 and D_3 is directly falling on photo diode in the receiver. The opamp (IC-1) has beam wired as a current to voltage converter. This voltage switches the Q_1 ON. Once Q_1 is ON, its collector is nearly at ground potential ($\cong 0.2$ V) which keeps Q_2 in the off-state with the result that the counter stays its initial setting.

When a light beam interrupted by Q_1 goes cut-off for a short time and Q_2 gets base drive through R_4 and R_5. Now Q_2 conducts again for a short time and the counter's solenoid gets a current pulse. Consequently, it increments by one count.

Try It Yourself

Try to comprehend the alignment process used in above project.

Project-

Electronic Switch Function

What do you know about working and assembly of electronic switches?

Introduction

An electronic switch is an electronic component or device that can switch an electrical circuit, interrupting the current or diverting it from one conductor to another. Typically, electronic switches use solid state devices such as transistors though vacuum tubes can be used as well in high voltage applications. In this experiment we learn the operation of electronic switches using electronic devices like transistors or SCRs. We have chosen these two as they are the most commonly used ones in this role, and in the electronic circuits of hobbyist's' interrerst.

Materials Required

- **Resistors** : All resistors of carbon film or carbon composition typ
  $R_1, R_2, R_3, R_4, R_5, R_6$: 22K, 1/4 W
  R_9, R_{10} : 3.3 K, 1/4W
  R_7 : 470Ω, 1/4W
  R_8 : 470Ω, 1/4W
- **Potentiometers**
  P_1 : 10K preset
- **Transistors**
  Q_1, Q_3 : 2N2222
  Q_2 : 2N2907
- **Diodes**
  D_1 : 1N4001
- **SCR**
  SCR-1 : SN100 or any other 200 volts SCR
  LED-1 to LED-4 : Red LEDs
- **Relay**
  RL-1 : 6 V DC relay
- **Miscellaneous**
  Battery : 9 V DC battery
  Switches SW_1 to SW_5 : ON/OFF switches

Assembly

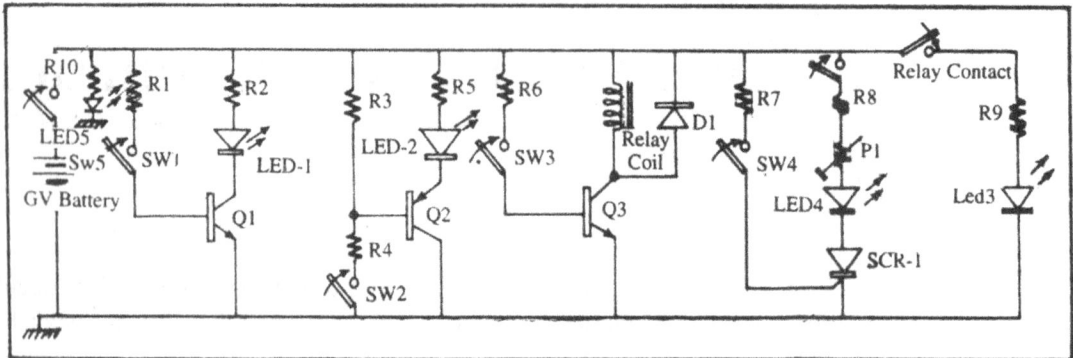

Fig. 11.1 : Circuit Diagram

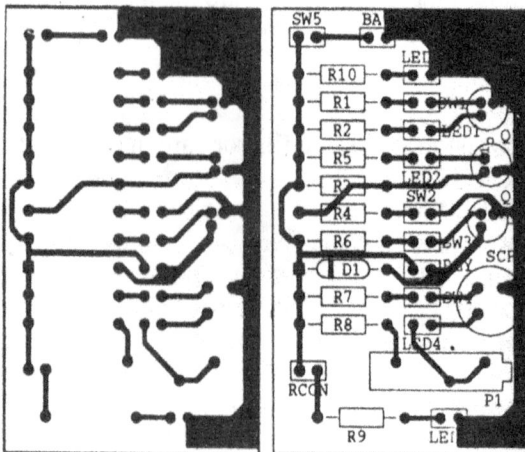

Fig. 11.2 : Circuit Diagram

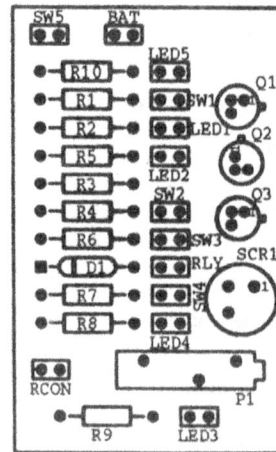

Fig. 11.3 : Circuit Diagram

The PCB layout as seen from components' side and components layout are shown in fig. 11.1, 11.2 respectively.

The first part of circuit fig. 11.3 starting from extreme left and comprising of resistors R_1, R_2 switches SW_2, LED_1 and transistor Q_1 illustrates the use of an NPN transistor as an ON/OFF type switch.

The second part consisting of resistors R_3, R_4, R_5 switch SW_2, LED_2 and transistor Q_2, depicts the use of a PNP and transistor as a switch.

SCR is another popular device used as a switch. However, it is a Latching device i.e. once triggered to the ON-state, it stays there even if the trigger is removed. Remember, in case of a transistor, you have to keep the switch SW_1 or SW_2 (as the case may be) closed in order to keep the transistor conducting and the LED glowing.

To Do and Notice

There are three different experiments you can perform with this simple circuit. These are:

   (i)   Use of NPN or PNP transistor as a switch

   (ii) Use of a transistor switch to drive a relay and

   (iii) Operation of an SCR as a latching switch

Before the start of every experiment, switch ON SW_5 to connect the battery voltage to the circuit. For the first experiment, switch on SW_1 and see if LED-1 glows. The LED should extinguish when you open SW_1. Similarly LED-2 would glow when SW_2 is closed.

Observe the energisation of the relay coil indicated by glowing LED-3 in the second experiment. In the third experiment, close switch SW_4, LED-4 glows. Now open switch SW_4, LED-4 continues to glow. Adjust the value of the potentiometer P_1 towards increasing resistance. You would observe the LED-4 getting extinguished at a certain stage. You will notice that you have to once again close the switch SW_4 and also adjust the present P_1 towards reducing resistance to light the LED-4. Again, the LED-4 continues to glow when the switch SW_4 is opened.

What Happens?

A transistor (PNP or NPN) is non-conducting or cut-off state (open circuit between its collector and emitter terminals) when its base-emitter junction is forward biased less than 0.6 V (base positive with repsect to emitter in NPN and emitter positive with respect to base in PNP transistors). When the base emitter junction's forward biased voltage is greater than 0.6V, the resulting base current drives the transistor ON and collector current equal to the base current multiplied by the current gain of the transistor flows in the collector emitter lead

In the SCR experiment shown, the moment you close the switch SW_4, SCR conducts and the LED-4 glows. Now the LED-4 can be observed to contiue to glow even if you open switch SW_4. There is only one method to bring the SCR back to the non-conducting state and that is to somehow bring its anode current below a certain minimum value which the SCR manufacturers refer to as the holding current. It is 10 mA for the choosen SCR.

Try It Yourself

Do you have understanding about mosfet switching function? If yes, try to establish.

Project-

Electronic Timer Clock

Can you make a time clock by using PIC6C54 micro-controller ?

Introduction

This electronic time clock project uses a 18 pin PIC16C54 microcontroller as its core in the display and setting of a simple 4 digit electronic clock. The four 7 segment displays are connected in multiplexing method and each segment is sequentially controlled by common cathode of each 7 segments.

Materials Required

18 Pin PIC 16 C54 micro-controller

Resistors	:	Four 10 kΩ, three 820 Ω, eight 100 Ω, four 4 kΩ
Transistors	:	4 PNP 2N5401 transistors
Display	:	4 seven segment display LEDs
Switch	:	3 switch SW_1, SW_2, SW_3, 1 relay switch
Battery	:	5V DC battery

Assembly

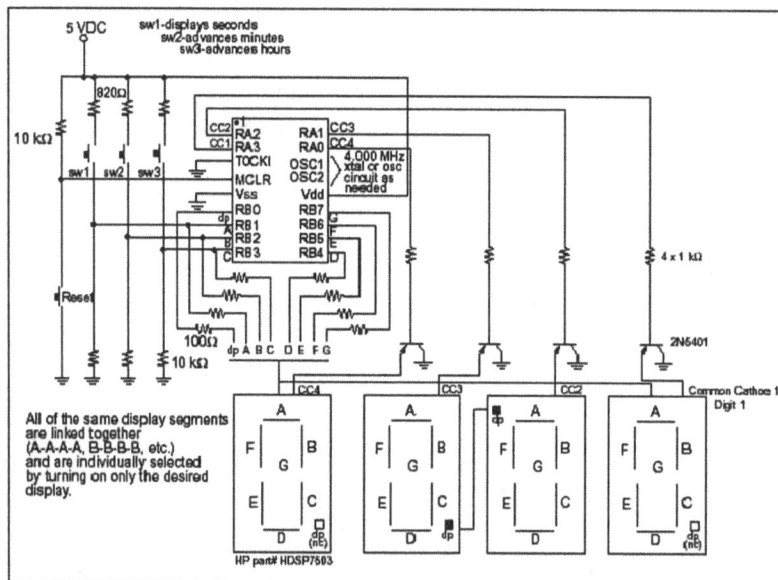

Fig. 12.1 : Circuit Diagram

The common cathodes are connected to 4 pins of Part A of the micro-controller. When the pin in Part A is low, the transistor connected to it will turn ON and that particular 7-segment is turns ON.

Part B will activate the segment of that particular 7-segment that has been chosen.

To Do and Notice

By multiplexing the selection of the 7-segment sequentially, an equal brightness of all the 7-segment display will be achieved.

What Happens?

Multiplexing helps to reduce the number of parts needed to connect to the 7 segments. It also keeps the cost of device low as fewer parts are needed. The set back is that the display may not be as bright as if it is driven directly without multiplexing.

Try It Yourself

Try to make an LED dimmer circuit using NE 555 timer.

Project-

Electronic Timer Switch

Can you make a timer switch which turns a lamp ON and OFF periodically ?

Introduction

A timer switch is a timer that operates an electric switch controlled by the timing mechanism. The timer may switch equipment ON, OFF, or both at a preset time after a preset interval or cyclically. This is a good project to simulate the presence of occupants in a house. These days when security is becoming a concern when no one is at home, having this device at home will deter the thief away from breaking in.

Materials Required

U_1	:	CD4060B binary counter
R_1	:	100 kΩ 1/4W, 5% carbon film resisto
R_2	:	2.2kΩ 1/4W, 1% carbon film resisto
R_3	:	1 M kΩ 1/4W, 1% carbon film resisto
R_4	:	3.9kΩ 1/4W, 5% carbon film resisto
R_5	:	10 kΩ 1/4W, 5% carbon film resisto
D_1, D_2, D_3, D_4	:	1N4148 diode
D_5	:	1N4003 diode
C_1	:	0.01 µF/25V ceramic capacitor
C_2	:	0.1 µF/25V ceramic capacitor
K_1	:	SPDT 12V relay with 5A/240 V AC contact rating
LAMP	:	Lamp

Assembly

The core of this electronic timer switch project uses a CD4060B binary counter. The binary counter has 10 outputs and the counter is counted by configuring the oscillator. Every negative clock will trigger the counter of the IC internally.

The timing of the circuit is affected by resistor R_3 and capacitor C_2. By connecting the four outputs in an AND configuration, the transistor Q_1 will only turn ON if all the 4 outputs are in logic '1'. If any of the logic is '0' the transistor will remain OFF.

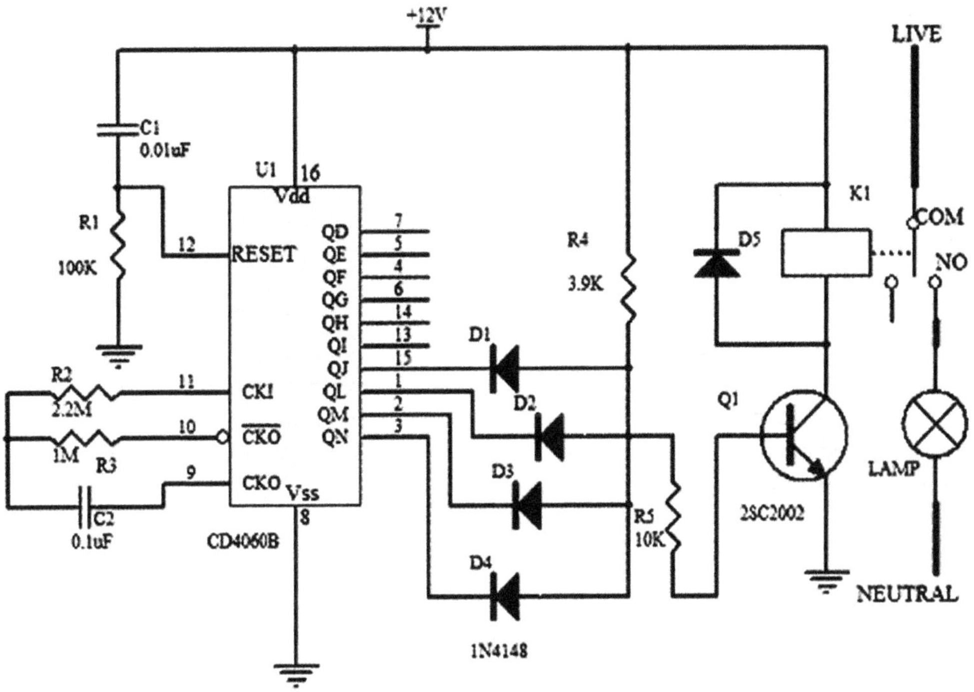

Fig. 13.1 : Circuit Diagram

To Do and Notice

When power up, after 60 minutes, the relay will turn ON for 100 seconds, OFF for the next 100 seconds, and ON again for 100 seconds before OFF again for next 60 minutes. This sequence will be repeated. A lamp which is connected to the relay will turn ON and OFF according to this timing.

What Happens?

For a complete cycle, the transistor will be ON twice when the output at pin 15, QJ goes to logic '1' and '0' twice when the other outputs QL, QM and QN remains at '1'. When this happens, the relay K_1 will switch status accordingly. The timing of the switching can be changed by changing the resistor values R_2, R_3 and C_2.

Note that since the oscillator is not using crystal, the timing may not be as accurate compared to the ideal calculation. In most cases, fine tuning the resistor and capacitor are good enough to make this project a success. To check whether the circuit is working, connect a LED in series with a 390 W resistor at output QD. It will flash ON and OFF as the oscillator oscillates.

Try It Yourself

Extend the OFF time in the above proejct.

Project-

Fire Alarm

Making a fire alarm syste

Introduction

Fire Alarm alerts us when there is a fire accident at home by ringing a siren sound. We are aware that there are many integrated circuits which can be used to generate the siren effect but we prefer to use basic elecronic components like resistors, capacitors and transistors to generate it so that you will clearly understand the internal working of it and it will be much useful for you as you will gain more knowledge by alayzing it instead of simply pre-designed integrated circuits.

Materials Required

TH-1	:	Thermistor
RV-2	:	Potentiometer arrangement
R_2	:	4.7 KΩ
R_1	:	120 Ω
R_5	:	10 Ω
R_3	:	4.7 KΩ
R_4	:	4.7 KΩ
R_6	:	56 KΩ
R_7	:	470 KΩ
R_8	:	22 KΩ
R_9	:	39 KΩ
R_{10}	:	47 KΩ
R_{12}	:	470 KΩ
Q_1, Q_2	:	BC547 transistor
Q_3	:	BC107 transistor
Q_4	:	2N2907 transistor
Q_5	:	2N2222 transistor
C_1	:	22μF
C_2	:	0.47

Battery :   9V batttery

Buzzer :   Microbuzzer

Assembly

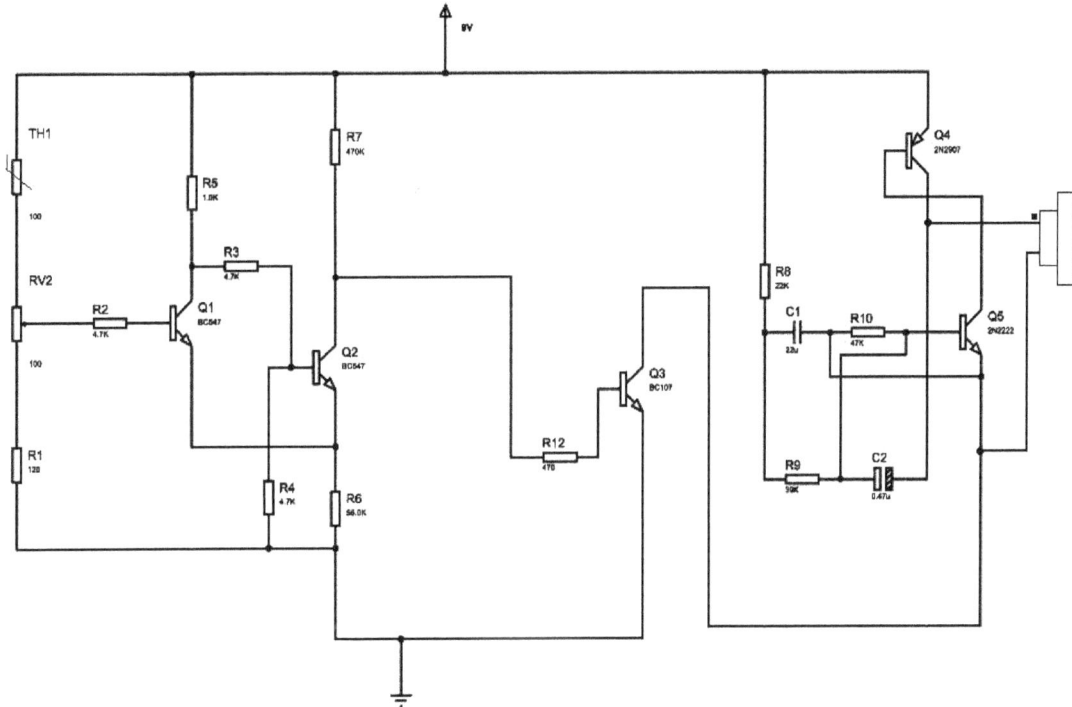

Fig. 14.1 : Circuit Diagram

This circuit uses a thermistor to sense the temperature. When it senses that the temperature of the environment increases above a given threshold, then it gives a signal. The temperature at which the circuit detects fire can be adjusted by using the potentiometer arrangement at V-2.

To Do and Notice

Suppose we set the temperature of thermistor at 50°C. When the temperature increases above the 50°C, the potentiometer arrangement produces a high voltage and siren starts. The temperature at which the circuit detects fire can be adjusted by using the potentiometer arrangment at V-2.

What Happens?

When the temperature increases above the set value, the potentiometer arrangement produces high voltage. This voltage is then given to BC 547 transistor in common emitter mode. When the base is given high input, it gets turned ON. When the transistor is turned ON, its collector voltage is reduced to low as the collector to emitter voltage decreases. The collector output voltage of the first transistor is given to the base as an input to the second BC547 NPN transistor. This transistor too in common emitter mode and as the input is low when the temperature threshold is reached, the output at the collector will rise high. In this state, it will turn ON the next transistor BC107. This transistor will now act as a switch for the siren circuit. This transistor can bear power quite larger than the BC 547 and it is also equipped with a heat sink for that purpose. When the BC107 turns ON, it allows current to pass from power supply to ground through collector thereby acting as an electronically controlled switch. When the current is passing, the siren circuit which is assembled as the load to the circuit is turned ON. Then you can hear the siren sound through the buzzer.

The capacitors used in the circuit are the main components in producing siren effect.

Try It Yourself
Try to design a Panic alarm circuit.

Project-

FM Jammer

How do you make a restricted area in which any FM radio can not work ?

Introduction

In old days when we used analog signal for communication, the Jamming circuit was easily used for producing high frequency noise signals, but today the trend is completely changed and the use of digital devices has taken place of analog devices. High frequency signals are not capable to block those signals from reaching the devices, so we need very high frequency signals to block actual signals from reaching the devices. The Jammer circuit produces high frequency noise signal which will confuse the receiver of a particular system from receiving the signal. User of the system feels that the circuit is not working properly.

Materials Required

R_1	:	15 KΩ resistor
R_2	:	3.9 KΩ resistor
R_3	:	220 KΩ resistor
C_1	:	Variable capacitor of 6 to 35pF
C_2	:	6 pF
C_3	:	0.01 pF
C_4	:	10 pF
Q_1	:	2N2222 transistor
L_1	:	Inductor
BT-1	:	9V battery
Antenna	:	Yagiuda antenna

Assembly

Capacitor C_1 and L_1 will constitute tank circuit. Transistor Q_1 will start the operaton of tank circuit. Resistors R_1 and R_2 will act as the biasing circuit and R_3 is used for limiting the emitter current in the circuit.

To Do and Notice

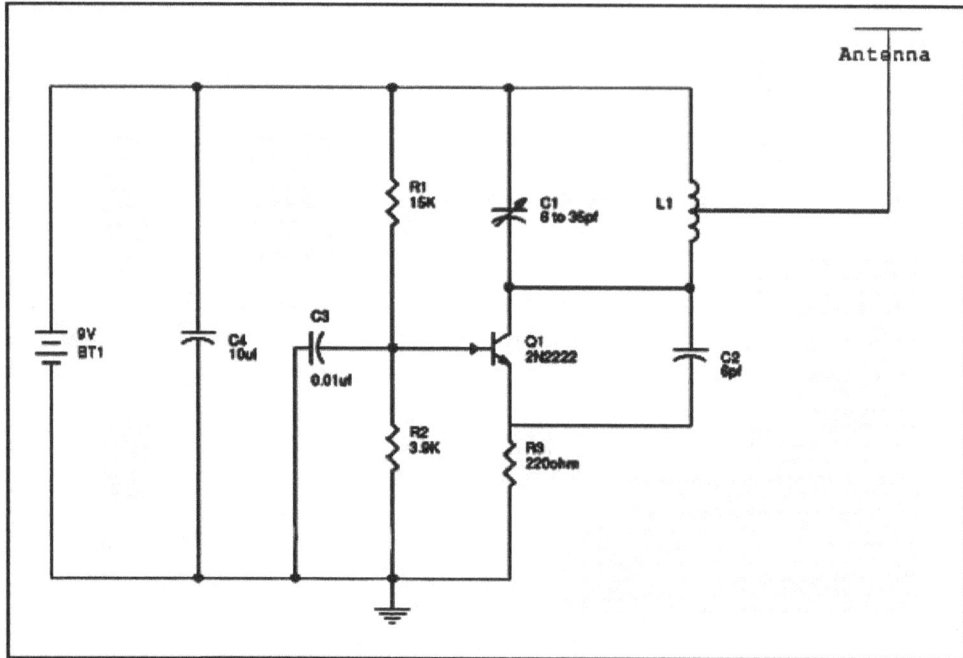

Fig. 15.1 : Circuit Diagram

When tank circuit is activated, it starts producing VHF (very high frequency signal) which jams the FM radio.

What Happens?

The Tank circuit will produce the high frequency signal, the capacitor C_1 is variable so that we can produce different frequency signals by adjusting the variable capacitor.

When Q_1 is turned ON, the tank circuit will start generating VHF signal which will create the noise in the original signal so that receiver can not receive the signal. Even if it is received also, the signal can not be used by the receiver circuit.

The value capacitor C_1 should be changed for every station for getting different frequency signals.

Different frequencies can be achieved by changing the values of capacitor and inductor by the formula

$$F = 1/(2 * Pi * sqrt (L * C))$$

Try It Yourself

Try to make a motion detector circuit.

Project-

Invisible Intruder Alarm

Can you make an intruder alarm for a locker which has audio as well as visible indicators?

Introduction

This is an ultra compact intruder alarm system based on the detection of an intrusion caused by the interruption of an infra-red light beam. This gadget has transmitter and receiver portions separately which is operated separately by 9V batteries. The transmitter and receiver gadgets can be mounted in an aligned position on the two facing walls of inside of a locker to give you audio (micro buzzer) and visible (LED) indication alarm. The LED can remotely be located away from the location of gadget also.

Materials Required

R_1	:	68Ω, 1W resistor
R_2	:	10MΩ, 1W resistor
R_3	:	1KΩ, 1/4W resistor
R_4	:	470Ω, 1/4W resistor
R_5, R_6	:	4.7KΩ, 1/4W resistor
R_7	:	100Ω, 1/4W resistor
D_1, D_6	:	IN4001 or equivalent diode
D_4	:	Photo diode, type No. TIL81
D_2, D_3	:	Infra-red emitting diode type No. TIL34
D_4	:	Infra-red emitting diode type No. TIL34
D_5, D_7	:	LEDs (D_7 red and D_5 green)
Q_1, Q_2	:	2N2222 transistors
SCR	:	OE101 or equivalent (any SCR with 100V breakdown voltage and 1A forward current rarings)
IC-1	:	LM324 (Quad opamp)
S_1, S_2	:	ON/OFF switches

Microbuzzer, two 9V batteries, 14-pin DIL IC socket.

Assembly

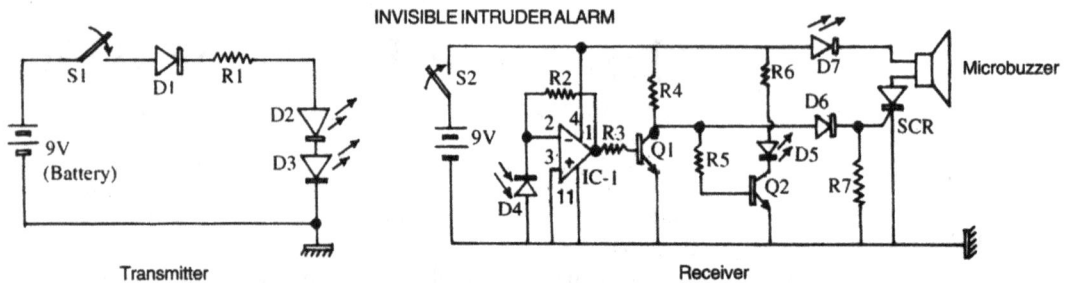

INVISIBLE INTRUDER ALARM

Fig. 16.1 : Circuit Diagram

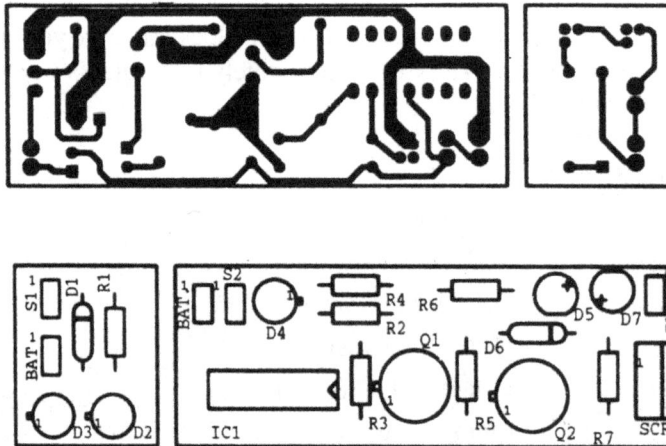

Fig. 16.2 : Circuit Diagram

The circuit shown in fig 16.1 circuit comprises a transmitter module and a receiver module. The transmitter module generates required current drive for the infra-red light emitting diode and the receiver module once aligned with the transmitter module senses any interruption of light beam caused by an intruder to activate the alarm.

Fig. 16.1 and Fig. 16.2 shows PCB layout and components layout diagrams respectively.

To Do and Notice

When there is an intrusion and the light beam is interrutped, LED (D_7) glows red and microbuzzer produces a sound. The SCR stays ON if once fired even if the gate drive is removed. Therefore, microbuzzzer continues sounding the alarm and LED (D_7) continues to glow even after the beam interruption is over and light beam path is restored.

Now, switch S_2 needs to be opened momentarily to reset the system and get it ready to detect the next intrusion.

What Happens?

In the transmitter, the current through the light emitting diode is being controlled by resistance R_1. The current here is about 60 mA.

In the receiver module, the infrared light directly fall on the photo diode in the receiver. The opamp (IC-1) converts the photo induced current into an equivalent voltage at the opamp output. This voltage switches transistor Q_1 ON. Once Q_1 is ON, its collector is at ground potential ($\cong 0.2V$) which keeps both the transistors Q_2 and SCR in OFF state with the result that LED D_7 does not glow and the microbuzzer too does not get the supply.

When there is an intrusion and the light beam is interrupted. Transistor Q_1 goes to cut-off. Transistor Q_2 conducts and also the SCR fires. LED ($_7$) glows red and microbuzzer produces a sound.

Try It Yourself

Comrehend and note down the role of LED (D_5) in the above project.

Project-

Metal Detector

How to design a circuit by which any explosive metals or illegal things like guns can be detected ?

Introduction

Metal dectctor is very common device for checking the person in shopping malls, hotels, cinema halls, to ensure that a person is not carrying any illegal explosive metals like guns, bombs etc. Metal detectors can be created easily and its circuit is not complex.

Materials Required

IC	:	TDA 0161
R_5, R_4, R_1	:	1 KΩ
R_2	:	330 Ω
R_3	:	120 Ω
RV-1	:	Variable resistor 10 kΩ
C_1, C_2	:	0.047 µF
BT-1	:	4V battery
SW-1	:	SPST key
Q_1	:	2N3904 transistor
Z-1	:	3V buzzer
D_1	:	LED
LC-circuit	:	680 PF

Assembly

As the fig. 17.1 shows the circuit diagram and fig 17.2 shows the block diagram. IC TDA0161 is a proximity sensor which can detect the objects without any physical interference. It works same as infrared sensor, proximity also releases a signal. It will not give output unless and until there is no change in the reflected back signal. If there is charge in signal, it will detect and give the output accordingly. LC circuit is a resonating circuit which will resonate when exact same frequency material comes near. The LC circuit consists of inductor and capacitor connected in parallel. When the capacitor is fully charged, the charge of the capacitor will be given to the inductor, here inductor will improve its magnetic field

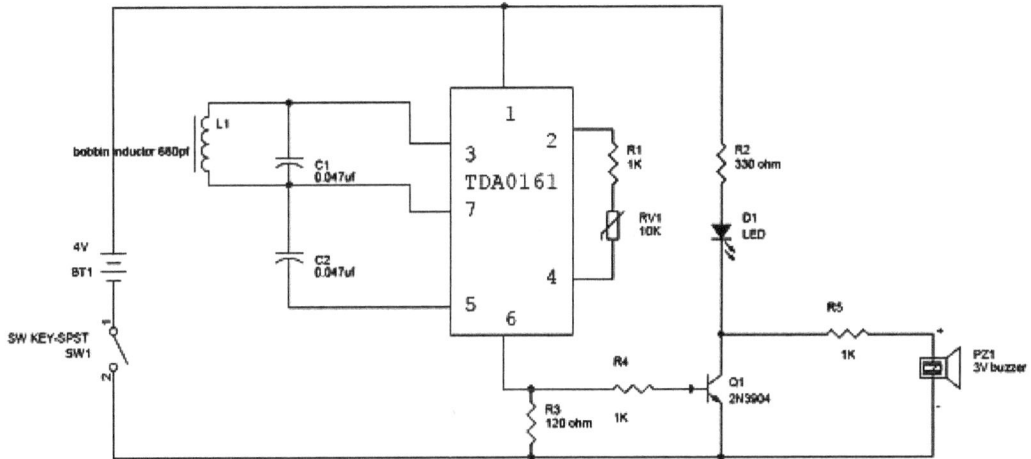

Fig. 17.1 : Circuit Diagram

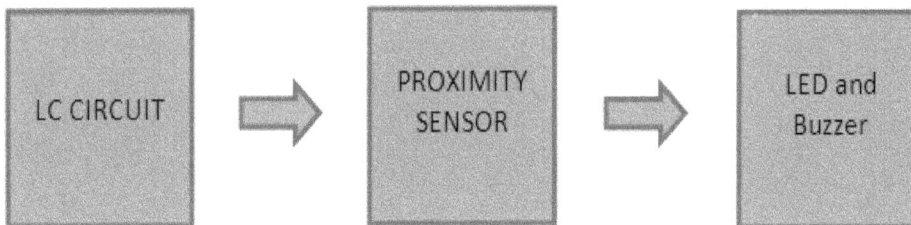

Fig. 17.2 : Block Diagram

To Do and Notice

When a metal is detected transistor Q_1 turns ON and LED glows and buzzer gives the buzz sound.

What Happens?

When the LC circuit that is L_1 and C_1 has got any resonating frequency from any metal which is near to it, the electric field will create and lead to induce current in the coil and changes in the signal flow through the coil. Variable resistor is used to change the proximity sensor value equal to the LC circuit, it is better to check the value when there is coil, not near to the metal. When the metal is detected the LC circuit will have changed signal. The charged signal is given to proximity detector, which will detect the charge in the signal and react accordinly. The output of the proximity sensor will be of a 1 mA when there is no metal detected, it will be around 10 mA when coil is near the metal.

When the output pin is high the resistor R_3 will provide voltage to transistor Q_1. Q_1 will turn ON and LED will glow and buzzer will give buzz. Resistor R_2 is used to limit the current flo .

Try It Yourself

Comprehend the working of proximity sensor.

Project-

Mobile Jammer

How to make a restricted area in which cell phones can not work ?

Introduction

By the use of a mobile Jammer circuit, we can make or cell phone restricted area. This circuit will work in the range of 100 meters i.e. it can block the signals of cells phones within 100 meters radius. This circuit can be used in TV transmission and also for remote controlled toys or play things. The usage of this type of circuits is banned in most of the countries.

Materials Required

R_1	:	39 kΩ
R_2	:	100 kΩ
C_1	:	15 pF
C_2	:	4.7 pF
C_3	:	4.7 pF
C_4	:	102 pF
C_5	:	1 μF
C_6	:	2.2 pF
C_7	:	10 pF
L_1	:	22 nH
Q_1	:	BF 494 transistor
Battery	:	3V battery
Antenna	:	Yagi uda antenna

Assembly

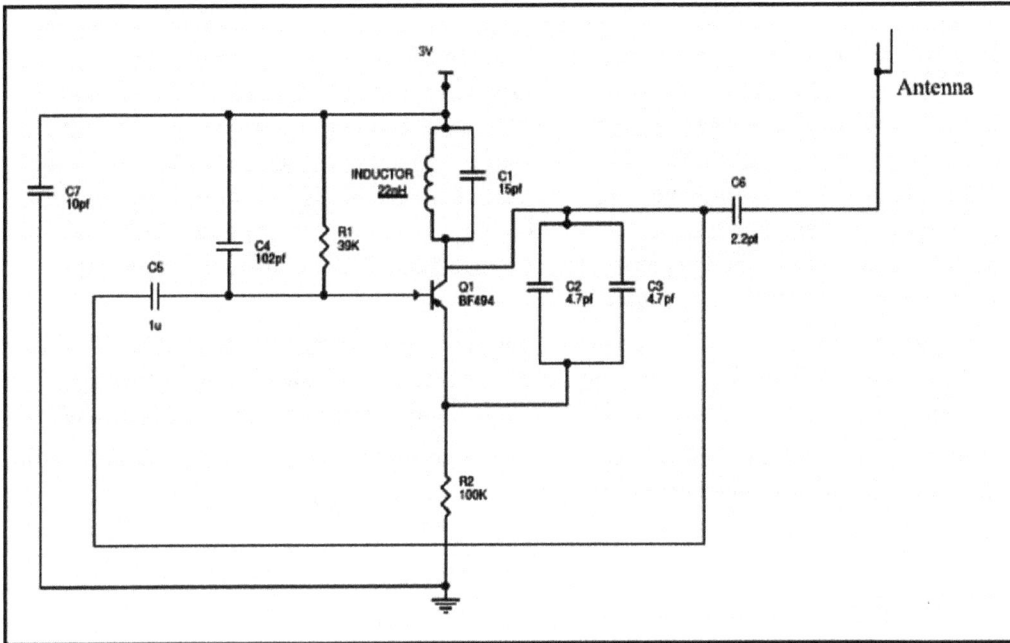

Fig. 18.1 : Circuit Diagram

The cell phone jammer circuit consists of following important circuits. When they are combined together, the output of that circuit will work as a jammer. The three circuits are :

   (i) RF amplifie

  (ii) Voltage controlled oscillator

 (iii) Tuning circuit

Here transistor Q_1, capacitor C_4 and C_5 and resistor R_1 constitute the RF amplifie .

Inductor L_1 and capacitor C_1 constitute of tuned circuit.

To Do and Notice

Cell phones work at frequency 450 MHz. To block this 450 MHz frequency, we also need to generate 450 MHz frequency with some noise which will act as simple blocking signal. This singal can block the cellphone signal from reaching the cell phones.

So, here in the above circuit, we generate 450 MHz frequency to block the actual cell phone signal. That's what the above circuit will act as a Jammer for blocking actual signal.

What Happens?

RF amplifier will amplify the signal generated by the tuned circuit. The amplication signal is given to the antenna through C_6 capacitor. Capacitor C_6 will remove the DC and allow only the

AC signal which is transmitted in the air. When the transistor Q_1 is turned ON, the tuned circuit at the collector will get turned ON. This tuned circuit will act as an oscillator with zero resistance. This tuned circuit will produce very high frequency with minimum damping. Both inductor and capacitor of tuned circuit will oscillate at its resonating frequency. The main function of capacitor is to store electric energy. The charged capacitor will allow the charge to flow through inductor. When the current flows across the inductor, it stores the magnetic energy. Now the voltage across the capacitor decreases, at some points complete magnetic energy is stored by inductor and the charge across capacitor will be zero. This process of capacitor, charging and energizing the inductor, repeats. After some time inductor will give charge to capacitor and become zero and they will oscillate and generate the frequency. RF amplifier feed is given through the capacitor C_5 to the collector terminal before C_6 gain signal.

The capacitors C_2 and C_3 are used for generating noise for the frequency generated by the tuned circuit. The feedback given by the RF amplifie , the frequency generated by the tuned circuit, the noise signal generated by C_2 and C_3 will be combined, amplified and transmitted to the ai .

If the circuit is not working, just increase the values of resistor and capacitors in the circuit. Also increase the frequency of tuned circuit by using this formula.

F = 1/[2 * Pi * sqrt (L * C)]

Try It Yourself
Try to build LED lamp dimmer circuit.

Project-

Morse Code

Do you know how a morse code sends and receives signals?

Introduction

The Morse code is a method of transmitting text information as a series of on-off tones, lights, or clicks that can be directly understood by a skilled listener or observer. The international morse encodes the ISO basic latin alphabet some extra latin letters, the arabic numerals and a small set of punctuation. Each character is represented by a unique sequence of dots and dashes. The duration of a dash is three times the duration of a dot. This is done by using 555 timer IC. This is one of the various ways for amature radio enthusiasts to practice the sending and receiving of morse code.

Material Required

U_1	:	NE 555 timer
R_1	:	10 kΩ 1/4W, 5% carbon film resisto
R_2	:	47 kΩ 1/4W, 5% carbon film resisto
VR_1	:	100 kΩ potentiometer
VR_2	:	10 kΩ potentiometer
C_1	:	0.01 µF/25V ceramic capacitor
E_1	:	33 µF/25V electrolytic capacitor
SPKR	:	8 Ω speaker
S_1	:	Switch
KEY	:	Key
BAT	:	9V Battery, Battery holder

Assembly

In this schematic, a 555 IC is used and configured as a timer astable mode. Once triggered, it will generate a frequency from its output at Pin 3. Once, a key is pressed, it will drive the 8Ω loudspeaker which is connected in parallel to potentiometer VR_2.

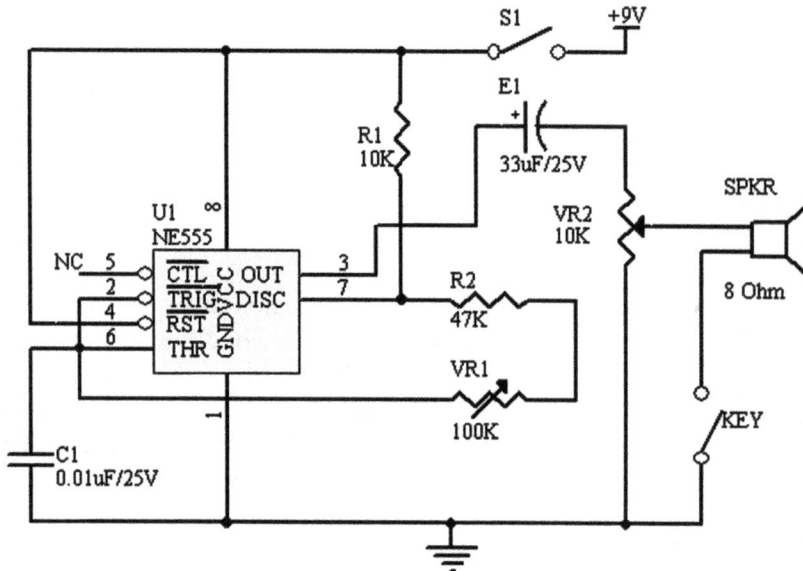

Fig. 19.1 : Circuit Diagram

To Do and Notice

The continous wave morse code uses a 555 timer which generates a *dit* or *dah* sound when the key is pressed.

What Happens?

The astable frequency of cricuit U_1 is given by the formula of 555 timer as shown below :

F (max) when VR_1 is set to 0 ohm

$$= \quad 1.44/[10 \text{ k}\Omega + 2 \ (47)] \ [0.01 \ \mu F]$$

$$= \quad 1.38 \text{ kHz.}$$

F (min) when VR_1 is set to 100 kW

$$= \quad 1.44/[10 \text{ k}\Omega + 2 \ (147 \text{ k}\Omega)] \ [0.01 \ \mu F]$$

$$= \quad 0.047 \text{ kHz.}$$

The frequency at the sound can be adjusted by varying the resistor of potentiometer VR_1. The volume of the speaker is adjusted by adjusting the potentiometer VR_2 which is connected is parallel with speaker and the key.

Try It Yourself

Try to make a Ham Radio.

Project-

Phone in Use

How will you determine the status of a phone line traffic by gl wing two LEDs of different colours?

Introduction

This project is known as phone in use which is a simple indicator that you can design and construct that displays the status of the phone line. If the line is in use, the yellow LED will turn ON. If it is not in use, the green LED will turn ON.

By having this indicator the user will not be interrupted by another user who wants to use the same line.

Materials Required

R_1	:	3.3 kΩ, 1/4W, 5% resistor
R_2	:	18 kΩ, 1/4W, 5% resistor
R_3	:	22 kΩ, 1/4W, 5% resistor
R_4	:	68 kΩ, 1/4W, 5% resistor
R_5	:	5.6 kΩ, 1/4W, 5% resistor
Q_1, Q_2	:	2N3393 NPN transistor
L_1	:	3 mm LED, yellow colour
L_2	:	3 mm LED, green colour
D_1, D_2, D_3, D_4	:	1N4002 diode

Assembly

As shown in the circuit, dioes D_1, D_2, D_3 and D_4 are used to ensure that in the event that the tip and ring of the line is reversed, the circuit can still be used.

When the telephone is in on hook condition, the voltage across the tip and ring is in the range of 48V DC to 50V. When it is in off-hook condition (the receiver is taken off its hook), the voltage drops to the range of 6V DC to 15V DC.

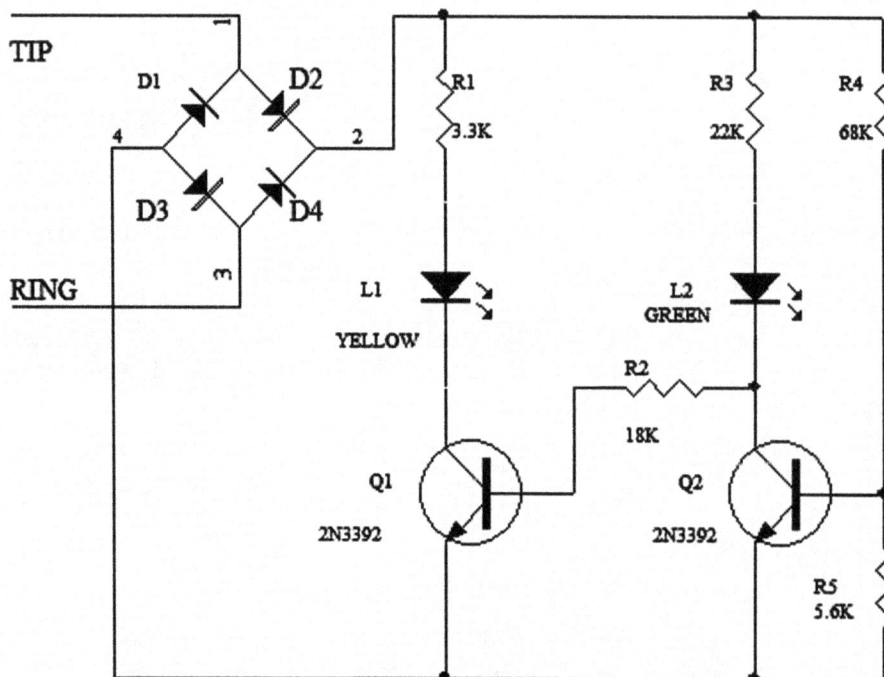

Fig. 20.1 : Circuit Diagram

To Do and Notice

When the telephone is on-hook condition, the green LED L_2 will slightly turn ON which indicates that the line is not in use. When the telephone is off hook condition, the yellow LED L_1 will turn ON.

What Happens?

When the telephones connected to the line which is on-hook condition, there is enough voltage to turn on transistor Q_2 through voltage divider R_4 and R_5. When Q_2 is ON, the green LED turns ON.

When the telephone goes to off-hook conditon, Q_2 will turn OFF. This allows current to flow through transistor Q_1 causing yellow LED turn ON.

Try It Yourself

Note down what happens when the phone is ringing ?

Project-

Railway Safety System

Can you make an automatic railway gate controller at the unmanned railway crossing with high speed alerting alarm system ?

Introduction

The main aim of this project is to operate and control unmanned railway crossing properly in order to avoid accidents. In a country like ours where there are many unmanned railway crossings, accidents are increasing day by day. These train accidents occur due to absence of human power in the railway. In order to overcome the accidents, the Railway Safety System is designed and installed at unmanned Railway crossing.

It has detectors at at a far distance on the railway track which helps us to know the arival and departure of a train. These detectors are attached to micro controller which activates motors and opens/closes the railway gate.

Another feature of this circuit is that it has an intelligent alerting system which detects the speed of the train that is arriving. If the speed is found to be higher than the normal speed, then micro controller automatically activates the alarm present at the gate.

Materials Required

10 kΩ variable resistors	:	4
100 kΩ variable resistors	:	4
2.2 kΩ variable resistors	:	1
101 Ω variable resistors	:	1
100 Ω variable resistors	:	1
10 kΩ variable resistors	:	1
BC 547 transistor	:	1
AC motor, 240 VAC	:	1
6 V DC Relay switch	:	1
Photodiodes	:	4
IR LEDs	:	4
ATMED A16L micro controller	:	1
470 μF/25V capacitors	:	2
10 μF/16V capacitors	:	2

7809 voltage regulator	:	1
7805 voltage regulator	:	1
Bridge rectifier	:	1
240V AC to 9V AC power adapter	:	1
230V-240V AC source	:	1

Assembly

Sensors
T- IR LED (Transmitter)
P-Photodiode(Receiver)

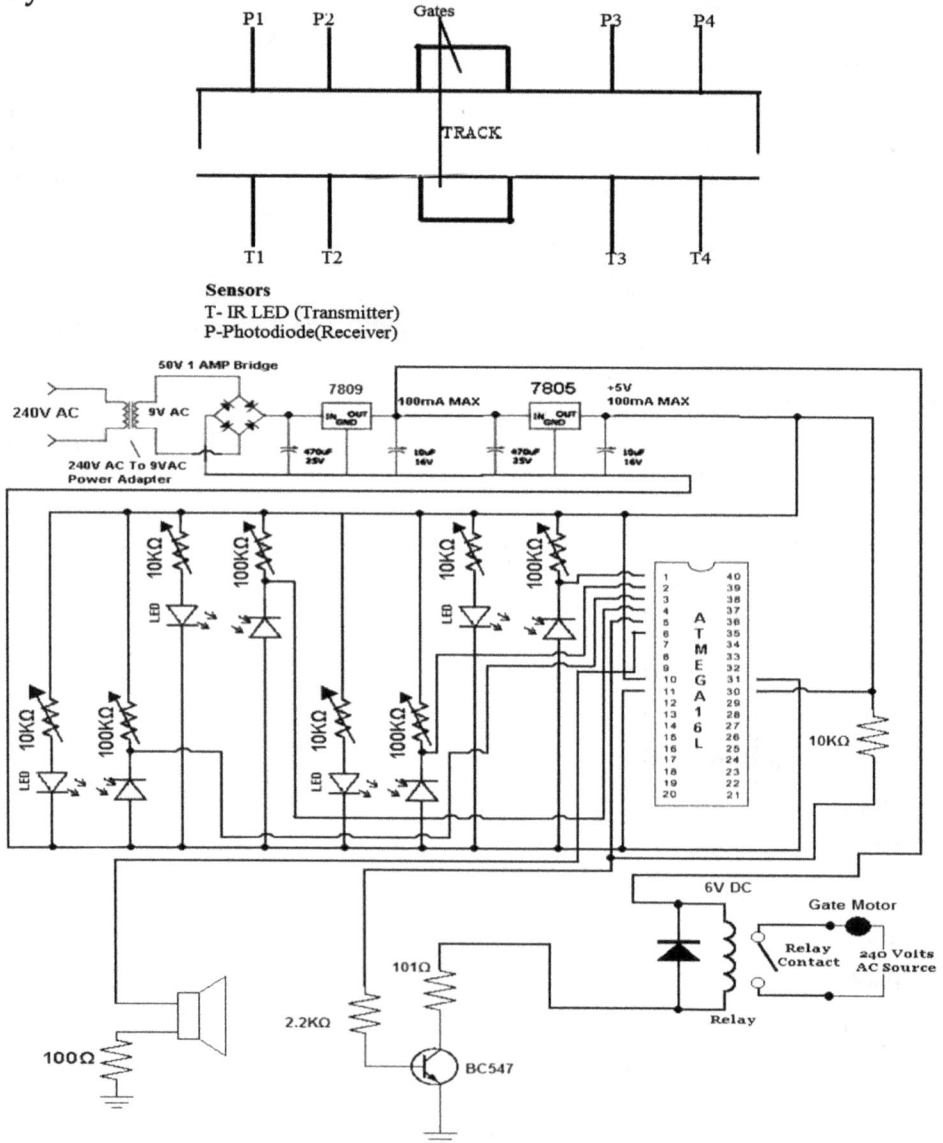

Fig. 21.1 : Circuit Diagram

The circuit consits of four IR LED-photodiode pairs arranged on either side of the gate such that IR as shown in the fig. 49.1. Initially, the transmitter is continously transmitting IR (infra red) light which falls on the receiver. When train arrives, it cuts the light falling on receiver.

To Do and Notice

Let us assume the train is arriving from left to right, now when the train cuts 1st sensor pair a counter is activated and when it crosses 2nd sensor pair the counter is stopped. This counter value gives the time period which is used to calculate the velocity of train. Sensor-2 output is sent to micro controller which makes the relay activate and causes the gate to be closed.

When last carriage of the train cuts sensor-4, microcontroller de-activates the relay and gates are opened.

What Happens?

The system basically comprises two IR LED-photodiode pairs which are installed on the railway track at about 1 meter apart, with transmitter and photodiode of the transmistter and the photodiode of each pair on the opposite sides of the track. The system displays the time taken by the train to cross this distance from one pair to the other with a resolution of 0.01 second from which the speed of the vehicle can be calculated as follows :

$$\text{Speed (kmph)} \quad = \quad \text{distance/time}$$

This circuit has been designed considering the maximum perimissible speed of trains as per the traffic rule. One more advantage of calculating the velocity of train is, if the speed of the train cross a limit i.e. it is travelling at an over speed, the passengers are alerted by the activated buzzer.

The microcontroller is used to process inputs that are provided by sensors and which generate the desired output appropriately.

Try It Yourself

Try to comprehend how a sensor detects the last carriage.

Project-

Rain Alarm

How to make a project which detects rain water and makes us alert ?

Introduction

Rain water detector will detect the rain and make an alert. This detector is used in irrigation fields, home automation, communication, automobiles etc.

Here is the simple and reliable circuit of rain water detector which can be constructed at low cost. Rain water sensor is the main component in the circuit. NE 555 timer is used to control the actions.

Materials Required

R_1, R_4	:	470 kΩ carbon film resisto
R_2	:	270 kΩ carbon film resisto
R_3	:	10 kΩ carbon film resisto
R_5, R_6	:	3.3 kΩ carbon film resisto
R_7	:	68 kΩ carbon film resisto
C_2, C_4	:	100 μF capacitor
C_1	:	220 μF/12V
C_3	:	0.01 μF
Q_1, Q_2, Q_3	:	BC148 transistor (NPN)
D_1	:	LED
D_2	:	1N4148 transistor
Battery	:	9V battery
U1	:	NE 555 timer IC
LS-1	:	Loud speaker 8Ω

Assembly

As seen in ig. 22.3 the rain sensor is the main component in the circuit. For the rain sensor, take the piece at bakelite or mica board and aluminum wire. Now, Aluminum wire should be posted on the lat board as shown in igure at rain sensor. This sensor is connected to the circuit i.e. between probes A and B.

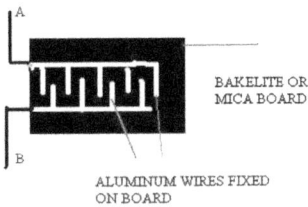

Fig. 22.1 : Water Sensor

Fig. 22.2 : Block Diagram

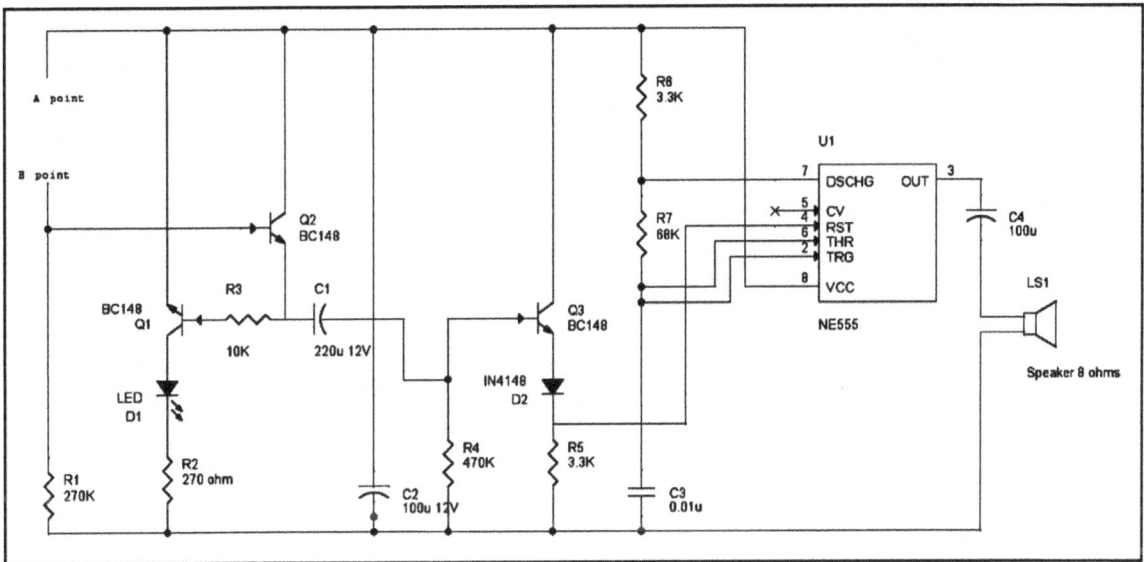

Fig. 22.3 : Circuit Diagram

Fig. 22.2 shows the block diagram of circuit. Fig 55.1 shows the main circuit diagram of our project.

To Do and Notice

If there is no rain, the resistance between the wires will be very high and there will be no conduction between the wires in the sensor. Its raining the water droplets will fall on the rain sensor which will decrease the resistance between the wires and wires on the sensor board will conduct and trigger NE 555 timer. Now loud speaker makes a sound. Hence, it leads to a loud sound.

What Happens?

When water will falls on the rain sensor, the aluminum wire develops resistance and gets conducted because of battery connector, the sensor and also to the circuit. When aluminum wires are connected, the transistor Q_1 will get turned ON and make LED to glow and also Q_2 will be turned ON as well. When the Q_2 is saturated, the capacitor C_1 will be shorted and make the transistor Q_3

to be turn ON. C_1 will get charged by the resistor R_4. The reset pin of 555 timer which is connected to the emitter of Q_3 will be made positive when Q_3 reaches the saturation mode.

Now Q_3 will generate the pulse at pin 3 and will make speaker ring an alarm. Capacitor C_4 will block the DC signal and allow only the variations in the signal which allows the speaker to generate sound. D_2 will not allow any reverse current from the timer because of R_4 and capacitors C_1. Q_3 will get cut-off after sometime and make the reset pin of 555 timer in negative and the speaker will stop making sound.

Try It Yourself
Make an LED lamp dimmer circuit.

Project-

Room Thermostat

Can you make a system which controls the ON and OFF of the chilled water valve based on the sensor temperature ?

Introduction

A thermostat is a component of a control system which senses the temperature of a system so that the temperature of system is maintained near a desired set point. The thermostat does this by switching heating or cooling devices ON or OFF, or regulating the flow of a heat transfer fluid as needed to maintain the correct temperature. Thermostats can be constructed in many ways and may use a variety of sensors to measure the temperature. This room thermostat project will focus on the chilled water control of a space that uses chilled water as its source of cooling. The controller basically consists of a comparator that controls the ON/OFF of the chilled water valve based on the sensor temperature.

Materials Required

RLY	:	SPDT 12V relay
U_1	:	LM358 operational amplifie
R_1	:	43 kΩ, 1/4W, 1% resistor
R_2	:	24 kΩ, 1/4W, 1% resistor
R_3	:	22 kΩ, 1/4W, 1% resistor
R_4	:	22 kΩ, 1/4W, 1% resistor
R_5	:	3.9 kΩ, 1/4W, 5% resistor
R_6	:	5.1 kΩ, 1/4W, 5% resistor
R_7	:	33 kΩ, 1/4W, 1% resistor
VR_1	:	5 kΩ potentiometer
VR_2	:	5 kΩ slide or rotary linear potentiometer
VD_1	:	Zener diode 5.6 V 0.25 W
Q_1	:	NPN transistor 2SC2002
Thermistor	:	Thermistor 20 kΩ @25°C, B > 74000 k
CW valve	:	Chilled water valve

Assembly

The circuit diagram shows the configuration of the room thermostat. The LM 358 opamp is used as a comparator to sense the inputs of the reference voltage (Pin 3) and room temperature (Pin 2). The thermistor used is a NTC (negative temperature coeffiecient) type where its resitance will drop when the temperature increases and vice-versa. It has a resistance of 20 kΩ at 25°C.

To Do and Notice

Suppose we set the room temperature at 25°C. When the room temeprature increases, turn ON the chilled water valve which allows the cold water to pass through the coil.

The cold air from the coil is transferred to the fan until the temperature of the room goes down to its set point again.

What Happens?

When the room temeprature increases, the thermistor resistance will drop and hence the output of the opamp will be high. This causes transistor Q_1 turn ON and the relay is turned ON. The composition of the relay contact will switch to NO (normally open) and ON the chilled water valve.

Fig 23.1 : Circuit Diagram

Try It Yourself

Make a project which controls the room temperature at constant desired level.

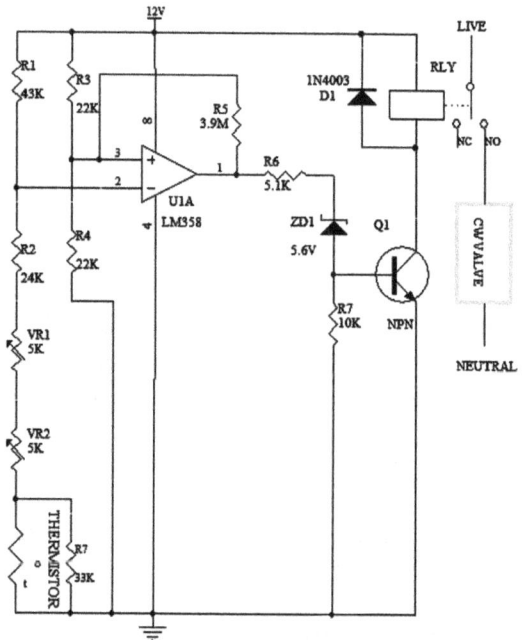

Project-

Sound Operated Light

Can you switch ON the lights with the sound of a clap?

Introduction

This gadget has a sensitivity adjust control so that the gadget can be made to operate at the intended sound level. Gadget getting triggered by a whisper and the gadget not responding to even a lound bang, both are undesirable situations.

The proposed gadget is designed to switch on and off alternately. That is if one sound pulse switches the light on, another one will switch it off. It is a desirable feature when you install this gadget with the sole aim of switching the lights ON and OFF at will without having to go to the switch board. Even if the intension is to keep the burglars away, alternate switching of lights could really frighten them away.

Materials Required

- **Resistors and capacitors**

R_1	:	10K, 1/4W
R_2	:	1K, 1/4W
R_3	:	470Ω, 1/4W
R_4	:	2.2K, 1/4W
P_1	:	1K preset
C_1	:	1000 µF, 50V (electrolytic)

- **Semiconductor**

D_1, D_2	:	1N4001 or equivalent
Q_1	:	2N2222
IC_1	:	741
IC_2	:	CD4027B

- **Components**

Transformer T_1	:	12V, 250 mA mains transformer condenser microphone
Relay	:	12V DC relay with at least one normally open contact
Fuse	:	0.5A with holder
Switch	:	S_1 (Main ON/OFF switch)

- **Miscellaneous**

Solder metal, multistrand wires, suitable cobinet etc.

Assembly

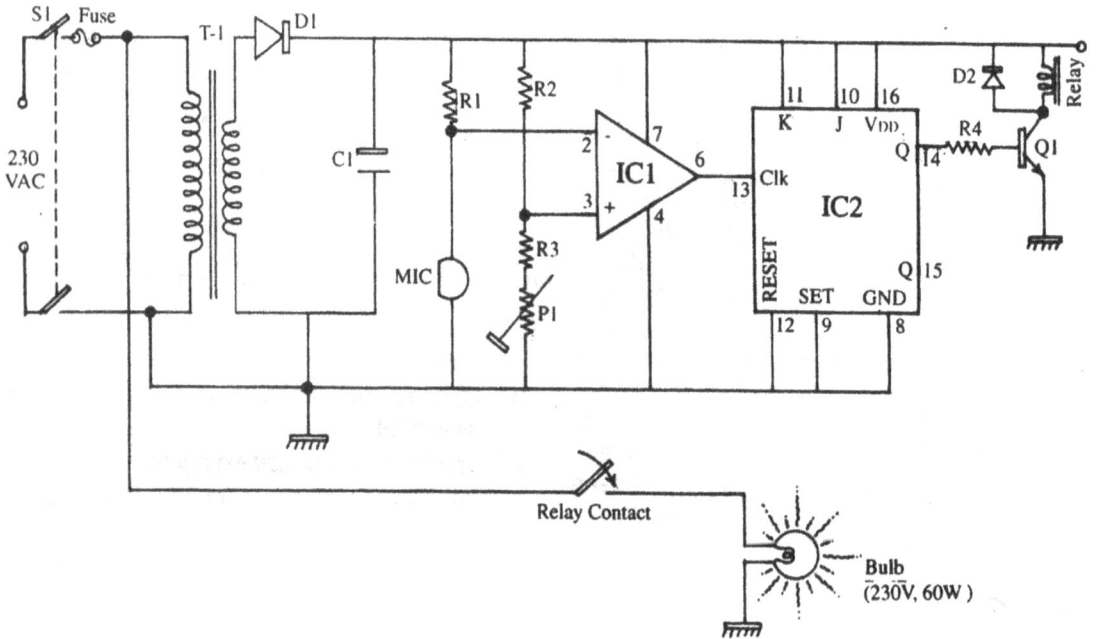

Fig. 24.1 : Circuit Diagram

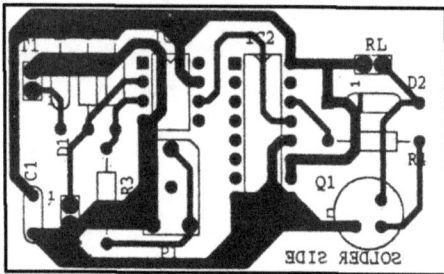

Fig. 24.2 : Circuit Diagram

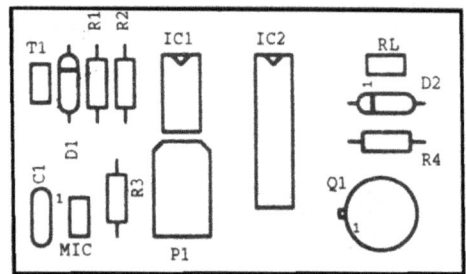

Fig. 24.3 : Circuit Diagram

Fig. 24.1 shows the circuit diagram of the gadget. Fig. 24.2 and Fig 24.3 show the PCB layout and components layout respectively.

The heart of the system is a J-K flip flop (IC_2) whose $(\overline{Q})$ output state is used to drive the relay coil. The flip-flo wired as a Toggle Flip-Flop in turn is clocked by the output of an opamp (IC_1) wired as a voltage comparator. The electric bulb gets AC input through relay contact.

Transformer T_1, diode D_1 and capacitor C_1 constitute the conventional AC/DC power supply generating +12V DC. The non-inverting input of the opamp is applied as a reference voltage produced by R_2, R_3 and the preset P_1. P_1 can be used to adjust the reference voltage.

To Do and Notice

The only critical parameter that is required to be carefully set and then verified in the sensitivity adjust. A higher value of reference voltage means a smaller sound level needed to force a change of state at the opamp output as in that case the voltage across the microphone needs to be changed by a smaller amount. A smaller reference voltage should be so adjusted (with the sensitivity adjust preset) as to produce the desired effect with intended sound level. Repeated tests should be carried out to establish the gadget reliability and that it does not trigger on false alarms.

What Happens?

The voltage at the inverting input is the same as that across the microphone. In the absence of any sound, this voltage is almost equal to the full DC voltage (= 12V DC) with the result that the output of opamp is initially low. The J-K flip-flo (IC$_2$) has been wired as a toggle flip-flo and its $\overline{Q}$ output is initially low. Transistor (Q$_1$) is in cut-off and the relay remains de-energised.

The AC power input connected to the bulb via relay contact thus does not reach the bulb and it remains extinghished.

In the presence of sound pulse current flows through the microphone and the voltage across the microphone goes down from +12V DC due to the potential divider arrangement formed by R$_1$ and the microphone. If the sound level is adequate so as to bring the voltage at the inverting input below the reference voltage at the non-inverting input, the opamp output goes high for a duration depending upon the duration of sound pulse. This positive-going pulse triggers the flip-flo and its toggles $\overline{Q}$. output goes high and the relay is energised with the result that relay's normally open contact closes and the bulb is lit. Another sound again toggles the flip-flo and the relay contact opens. The bulb extinguishes again. Thus the bulb lights up and extinguishes alternately if there are re-current sound pulses.

Try It Yourself

Design a true burglar's alarm that operates automatically taking the up sounds like those of opening of a door.

❐

Project-

Thermostat

How to make an automatic room temperature control system?

Introduction

This project focuses on the heating control of a space that uses electric heater as its source of heating. It basically consists of a comparator that controls the ON and OFF of the electric heater based on the sensor temperature. A thermostat is used to control and regulate the temperature of a space to its desired set point. They are used entensively in HVACR which stands for heating, ventilation, air conditioning and refrigeration.

Materials Required

RLY	:	SPDT 12V relay
U_1	:	LM 358 operational amplifie
R_1	:	43 kΩ 1/4W, 1% resistor
R_2	:	24 kΩ 1/4W, 1% resistor
R_3	:	22 kΩ 1/4W, 1% resistor
R_4	:	22 kΩ 1/4W, 1% resistor
R_5	:	3.9 mΩ 1/4W, 5% resistor
R_6	:	5.1 kΩ 1/4W, 5% resistor
R_7	:	33 kΩ 1/4W, 1% resistor
VR_1	:	5 kΩ potentiometer
VR_2	:	5 kΩ slide or rotary liner potentiometer
ZD_1	:	Zener diode 5.6V 0.25 W
Q_1	:	NPN transistor 2SC 2002
Thermistor	:	Termister 20 kΩ @ 25°C, B > 4000kW
HTR	:	Heater

Assembly

The circuit diagram in fig 64.1 shows the configuration of the thermostat. The LM 358 opamp is used as a comparator to sense inputs of the reference voltage (PIN-3) and room temperature (PIN-2).

The thermistor used is a NTC (negagive temeprature coefficient) type where its resistance will drop when the temperature increases and vice-versa.

To Do and Notice

Suppose we set the room temperature at 25°C.

When the room temperature drops, the relay to turn off and heater will conduct until the temperature of the room rises again.

What Happened?

The used thermistor has a resistance 20 kΩ at 25°C. When room temperature drops, the thermistor resistance will go up and hence the output of the operational amplifier   ill be low. This causes the relay to turn off and the heater will conduct until the temperature of the room rises again.

Fig 25.1 : Circuit Diagram

The circuit is calibrated using variable resistor VR_1. Set the lever of the slide potentiometer or rotary potentiometer VR_2 to 25°C location.

Place the thermistor at a space where the temperature is at 25°C. By varying VR_1, set the resistance at the position between the ON and OFF of the relay. Use a suitable contact relay rating according to the load of the heater.

Try It Yourself

Comprend how a refrigrator cools inside below the room temperature.

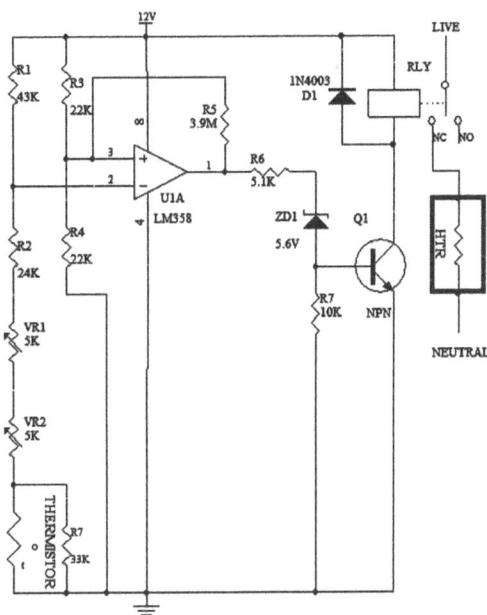

Project-

Time Delay Circuit

Can you make a simple time delay circuit ?

Introduction

In the design of analog circuits, there are times when you will need to delay a pulse that comes into a circuit before being used for the next process. This circuit uses a 555 timer to delay a pulse that comes into a maximum time of 75 seconds.

The timing of delay can also be changed by changing the resistor value of VR_1 and the capacitor value of E based on the time delay formula of $t = 0.69 \, RC$.

Materials Required

U_1	:	555 timer
R_1	:	470 kΩ 1/4 W 5% carbon film resisto
VR_1	:	5 MΩ variable resistor
Q_1	:	BC 556 PNP transistor
E	:	22 μF/25V electrolytic capacitor

Assembly

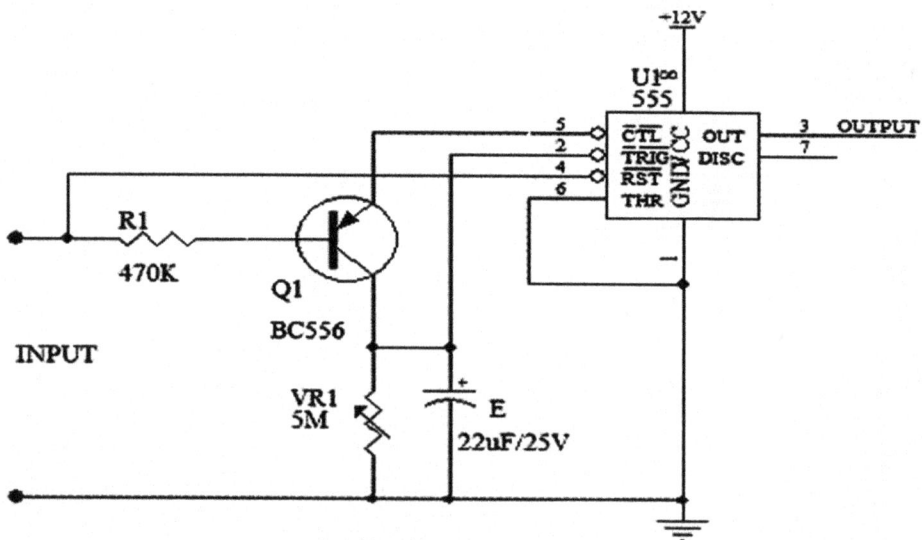

Fig. 26.1 : Circuit Diagram

In order for the output to go high, the reset pin of 555 timer (pin 4) must be high and the trigger pin (pin-2) voltage level must be below a third of the level of the power supply to the IC. When there is no pulse being applied to the input, transistor Q_1 will turn ON and capacitor E will be charged.

To Do and Notice

Once a pulse is applied to the input, capacitor E_1 will be discharged through VR_1.

This discharging time will be the delay time for the pulse. After whole charge discharges the transisor Q_1 will turn ON.

What Happens?

Once a pulse is applied to the input, transistor Q_1 will turn OFF and pin 4 reset pin is held high. This causes the capacitor E_1 to be discharged through VR_1 resistor. The time delay will depend on the discharge of capacitor E to a third of the supply before output of 555 goes high by changing different values of VR_1 and E to get different time delay.

If the maximum value of potentiometer is set to 5 MΩ, the time delay of the pulse will be 75 seconds.

Try It Yourself

Explain the working of 555 timer in monostable and astable mode.

Project-

Wireless Switch

Can you design a circuit by which any electrical appliance make ON/OFF without physical contact with the appliance ?

Introduction

Generally, appliances that we use in our home are being controlled with the help of devices like switches, sensors. However sometimes it is dangerous to have physical contact with these switches. So, to overcome these dangers, we have explained a circuit that needs no physical contact with the appliance. In this circuit all you need to pass is your hand above LDR. As you first pass your hand over LDR, the device connected with it starts and remains in that state till you again pass your hand above LDR.

Materials Required

IC_2	:	CD4027, J-K flip fl
IC_1	:	LM741 opamp
R_1	:	1KΩ, carbon film resisto
R_2	:	82Ω, carbon film resisto
R_3	:	10Ω, carbon film resisto
VR-1	:	5MΩ, variable resistor
Relay	:	Relay switch
D_1	:	1N4001 diode
LDR	:	LDR
T_1	:	BC547 transistor
C_1	:	0.1 μF
Battery	:	6V battery

Assembly

This circuit mainly depends on two ICs. First one is LM741 which is an opamp amplifie . This IC is used to increase the voltage level at the output hundred of a thousand times as comapred with the input. Another IC is most generally used J-K flip flop. This IC works in toggle mode and is based on the J-K flip flop master slave concept

It is used to alter the state when the signal is given to the any one of the input terminals and can get more than single output. These are four input pins J-K flip flop named J and K along with set and reset pin and Q and $\overline{Q}$ for output.

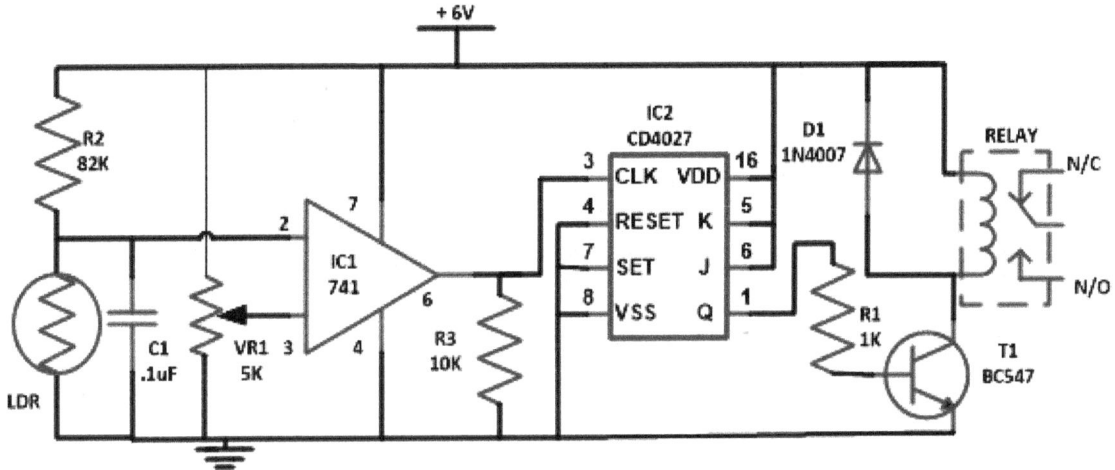

Fig. 27.1 : Circuit Diagram

To Do and Notice

Now, as soon as someone passes their hand over LDR the circuit start it remains start until again pass your hand over the LDR.

What Happens?

When you pass your hand over LDR IC_1, pin-3 is set to high as compared with the IC_1 pin-2 as a result of its pin-6 reaches to high state which supplies a clock pulse to 13 pins at IC_2. The logic level is present at the input terminal i.e. J and K guarded the state of a flip-flo with the help of some internal control. With positive-going cycle of the clock the changes occur. Set and Reset pin are not dependent on the clock and it starts when high signal is provied to any one at the input pins.

The circuit is triggered on the primary rim of the switch pulse and output vary when you once more place your hand over LDR. As in the circuit you can find 5 and K both attached to high input which implies that at each transition of clock pulse whether negative or positive pin 13 fluctuates between high and low. This can be demonstrated with help of JK flip flop truth table. Hence, when it gets the pulse from clock from IC_1 as of hand above LDR, transistor attached to pin 15 begins and output is get with the help of relay attached in the circuit. With the help of VR_1 sensitivity can be adjusted.

Try It Yourself

Try to make a function generator which can be used in different ways to produce signals ?

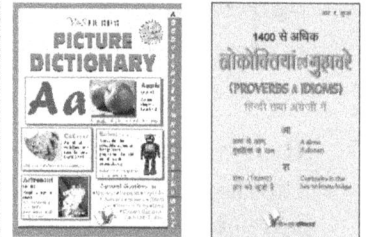

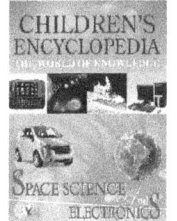

PARENTING

FAMILY & RELATIONS

Contact us at sales@vspublishers.com

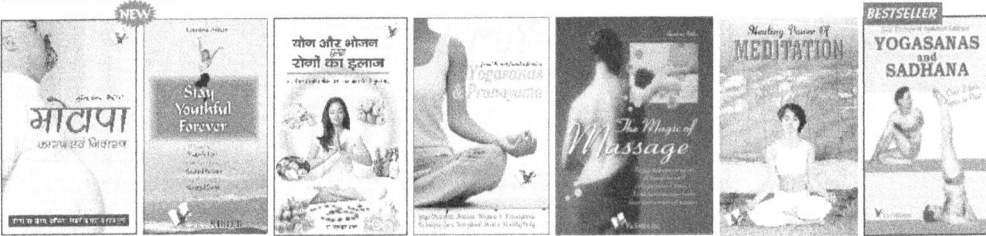

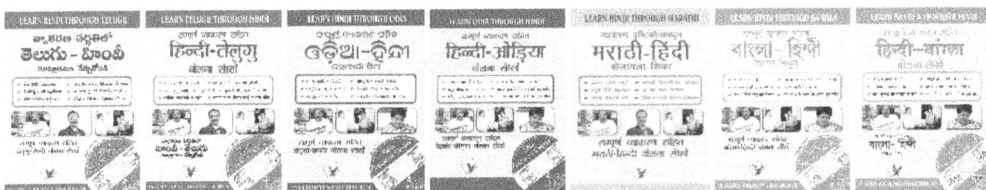

CONCISE DICTIONARIES

ACADEMIC BOOKS

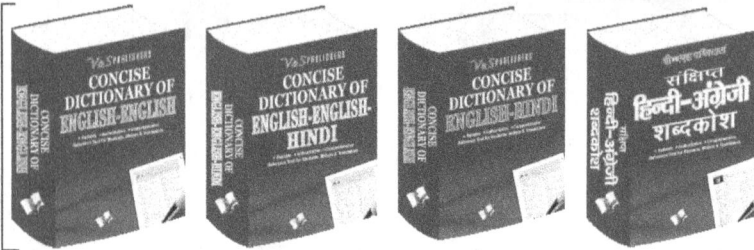

CONCISE DICTIONARY OF **ENGLISH-ENGLISH**

CONCISE DICTIONARY OF **ENGLISH-ENGLISH-HINDI**

CONCISE DICTIONARY OF **ENGLISH-HINDI**

संक्षिप्त **हिन्दी-अंग्रेजी** शब्दकोश

ENGLISH DICTIONARIES

HINDI DICTIONARIES

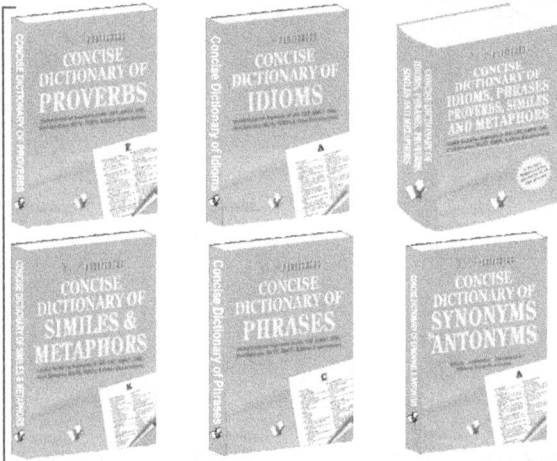

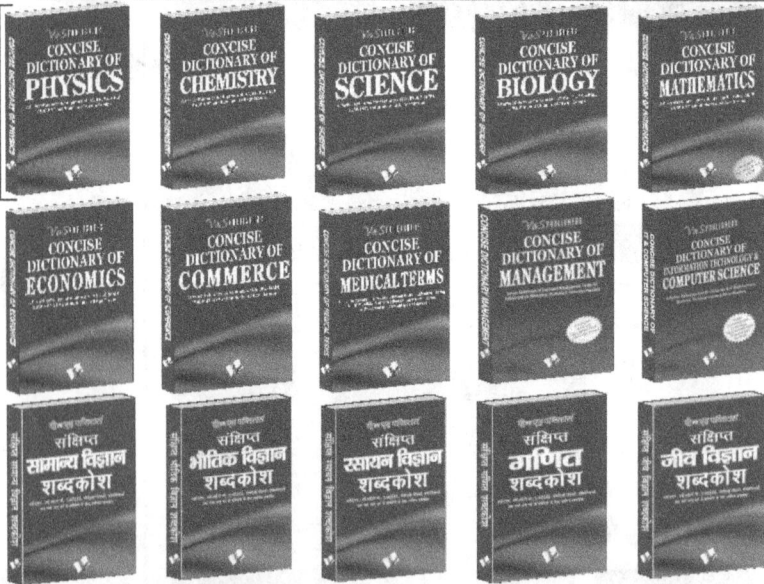

CONCISE DICTIONARY OF **PROVERBS**

CONCISE DICTIONARY OF **IDIOMS**

CONCISE DICTIONARY OF **IDIOMS, PHRASES, PROVERBS, SIMILES AND METAPHORS**

संक्षिप्त **पर्यायवाची** शब्दकोश

CONCISE DICTIONARY OF **SIMILES & METAPHORS**

CONCISE DICTIONARY OF **PHRASES**

CONCISE DICTIONARY OF **SYNONYMS ANTONYMS**

संक्षिप्त **विलोम** शब्दकोश

ENGLISH DICTIOANRIES

CONCISE DICTIONARY OF **PHYSICS**

CONCISE DICTIONARY OF **CHEMISTRY**

CONCISE DICTIONARY OF **SCIENCE**

CONCISE DICTIONARY OF **BIOLOGY**

CONCISE DICTIONARY OF **MATHEMATICS**

CONCISE DICTIONARY OF **ECONOMICS**

CONCISE DICTIONARY OF **COMMERCE**

CONCISE DICTIONARY OF **MEDICAL TERMS**

CONCISE DICTIONARY OF **MANAGEMENT**

CONCISE DICTIONARY OF **INFORMATION TECHNOLOGY & COMPUTER SCIENCE**

संक्षिप्त **सामान्य विज्ञान** शब्दकोश

संक्षिप्त **भौतिक विज्ञान** शब्दकोश

संक्षिप्त **रसायन विज्ञान** शब्दकोश

संक्षिप्त **गणित** शब्दकोश

संक्षिप्त **जीव विज्ञान** शब्दकोश

CAT 2015 Common Admission Test

CAT 2015

General Studies Paper I 2015

General Studies Paper II 2015

General Studies Solved & Practice Papers

Mathematics Formulae

Civil Services PLANNER

NTSE MAT + SAT

General Knowledge 2015

सामान्य ज्ञान

CRASH COURSE **JEE (MAIN)/AIEEE PHYSICS**

CRASH COURSE **JEE (MAIN)/AIEEE CHEMISTRY**

CRASH COURSE **JEE (MAIN)/AIEEE MATHEMATICS**

MATHS Olympiad for IMO Aspirants

FICTION

Contact us at sales@vspublishers.com

www.ingramcontent.com/pod-product-compliance
Lightning Source LLC
Chambersburg PA
CBHW081421270326
41931CB00015B/3356